Environmental Engineering

Series Editors: U. Förstner, R. J. Murphy, W. H. Rulkens

Dieter Vogelsang

Environmental Geophysics

A Practical Guide

With 113 Figures

Springer-Verlag
Berlin Heidelberg NewYork
London Paris Tokyo
HongKong Barcelona Budapest

Series Editors

Prof. Dr. U. Förstner

Arbeitsbereich Umweltschutztechnik
Technische Universität Hamburg-Harburg
Eißendorfer Straße 40
D-21073 Hamburg, Germany

Prof. Robert J. Murphy

Dept. of Civil Engineering and Mechanics
College of Engineering
University of South Florida
4202 East Fowler Avenue, ENG 118
Tampa, FL 33620-5350, USA

Prof. Dr. ir. W. H. Rulkens

Wageningen Agricultural University
Dept. of Environmental Technology
Bomenweg 2, P.O. Box 8129
NL-6700 EV Wageningen, The Netherlands

Author

Professor
Dr. rer. nat. Dieter Vogelsang
Kampstraße 70
D-30629 Hannover
Germany

ISBN-13:978-3-642-85143-8   e-ISBN-13:978-3-642-85141-4
DOI: 10.1007/978-3-642-85141-4

CIP-data applied for

Typesetting: Data-conversion by Fototsatz-Service Köhler OHG, Würzburg
SPIN:10085882   61/3020-5 4 3 2 1 0 - Printed on acid-free paper

Preface

Applied geophysics was developed to explore the raw materials required by civilization. This book defines its new environmental task: to investigate the extent and nature of buried contaminated waste and leachates. It describes the possibilities, advantages and shortcomings of geophysics in detail and in plain words, without referring to mathematical formulaes or scientific jargon. *Environmental Geophysics* may also serve as a simple introduction to geophysics for students inexperienced in mathematics.

Firstly, geophysical methods are described. Later, more than 80 environmental case histories from the USA and abroad are discussed and documented with 113 figures. The last three chapters present the gist of the book through condensed, lucid tables for the hurried reader. This briefing comprises cost estimates for geophysical surveys and offers advice for the proper choice of methods and for the compilation of tenders.

This book will enable engineers, scientists and lawyers to appreciate the great possibilities of geophysics in the assessment of environmental risks. This new branch of science allows continuous spatial coverage, considerable cost and time saving, is noninvasive and guarantees high standards of industrial safety.

This American edition is based roughly on the second edition of *Geophysik an Altlasten*. The German book was completely revised and considerably enlarged to meet the high American standards and requirements. The author would like to thank Mrs. Barbara Eder for her help with the translation.

Hannover, March 1994 Dieter Vogelsang

Contents

1 Introduction

1.1 Targets

This book was written to show how geophysics might be applied to the solution of environmental problems. Investigations of ground water, soil and rocks are described the subject of air pollution is not addressed, however.

Non-geophysicists who deal with environmental problems, like engineers, scientists of other faculties, lawyers and community personnel, are introduced to the application of geophysical methods. Complicated scientific elaborations and mathematical formulaes are omitted. To contribute to a better understanding of the practical feasibility and validity of these methods, many illustrated case histories are presented.

Additionally, examples of tenders for geophysical surveys are included. Cost-benefit analyses compare geophysics, drilling and probing programs and demonstrate the economic advantages of using geophysical measurements to solve environmental problems.

1.2 Fundamentals

Geophysical surveys of hazardous waste deposits are rarely published, since they contain the classified data of companies, government offices and corporations. Because of this, the high success rate of geophysical activities at contaminated sites is as of yet known only to insiders.

Thanks to an intiative of the German state of Baden-Wuerttemberg, which allowed a comparison of the results of 60 geophysical surveys at eight model sites at hazardous waste dumps, essential high-quality information became available.

Geophysical results published by scientists from Europe and America plus investigations carried out by the author have enlarged the scope of this book considerably.

1.3 Preconditions

Over the past 75 years or so, geophysical methods have been developed, mainly for the prospecting of deep-seated deposits of hydrocarbons and ores. During this time, many instruments have been constructed and extensive experience in evaluating and interpreting data has been accumulated.

To cope with increasing rates of consumption of raw materials, the geophysical exploration of the resources of fossil energy and ore has had to extend to greater depths. Geophysics has thus helped to satisfy the raw material needs of our high-tech civilization.

However, the task of controlling the disposal of the voluminous remnants of used raw materials is a new one for geophysics. It necessitates a completely new orientation. Now, the hitherto eliminated surface effects that have been suppressing the signals of deep-seated deposits have become a new subject of investigation.

There is no need to develop completely new geophysical methods for the exploration of hazardous sites. The above mentioned instruments and experiences can be utilized fully. However, it is necessary to adapt the various methods to the new environmental problems. It is, for instance, necessary to lay out very fine meshed grids, which allow high accuracy in shallow depths.

Since geophysical evaluations depend strongly on the geological and hydrogeological structures of the ground, it is necessary to consider these when any geological interpretation is derived from geophysical data.

1.4 Cooperation

Difficulties and misunderstandings may arise between geophysicists and engineers due to the lack of mutual understanding. Good cooperation is essential, since engineers control and finance most geophysical surveys.

The reason for this is mostly a different consideration of geophysical results. While engineers regard every figure as "absolute", geophysical data may be relative, though they are based on the exact sciences of mathematics and physics.

This paradoxical phenomenon comes from the great variety and complexity of the physical properties of geophysical targets. Their physical parameters must often be smoothed and averaged before they can be treated by mathematical formulaes.

Table 1.1 elucidates this dilemma. Furthermore, advice is included that engineers should describe their environmental problems as exactly as possible, so that the most suitable array may be chosen.

The geophysicist should, in return, inform the engineer precisely about the limitations and restrictions of geophysical interpretation and should point out the margins of error.

Further misunderstandings may arise from the presentation of data. They should be readable and understandable, even for the geophysical layman. Most impressive are certainly three-dimensional colored maps or sections. They portray many details, and make complicated issues understandable.

Columns of numbers, pseudosections and contour maps are less comprehensible. But it must be considered that a "beautiful" picture can rarely be used for reinterpretation or upgrading, since it is mostly impossible to obtain precise data from three-dimensional and/or color illustrations.

Table 1.1. Assessment of geophysical results

Engineers	Geophysicists
Absolutely unerring	Ambiguous, several equivalent solutions possible
All properties are exactly expressed in measuring units	Properties can be determined only approximately
No interpretation is neccessary	Interpretation is essential

Clear and correct description of problems ⟶

⟵ Straightforward description of restrictions

It is best to combine the beautiful with the exact: geophysical reports should contain the readable original data, perhaps on data carriers like PC discs or written lists. Additionally, colored pseudosections, 3D projections etc. should be included.

Color has to be used with precaution: the nuances must agree with the data-steps. Differences in colors are difficult to see in yellow or green, but provide good contrasts in red or blue. The introduction of colours means also the opening of an additional dimension for presentation; by colouring, one more set of data can be shown in a drawing.

Depth data for engineers should be in meters or feet only. The tendency of geophysicists to divide the y-axes for the vertical extension in ms, ns, mV, nT or other physical units may be misleading because these units may be considered equivalent to depth measures in meters. If it is not possible to reveal the actual depth, the y-axes should not be marked at all. Naturally, the reason for this has to be mentioned.

Overestimations of depth penetration may occur, especially in interpretations of electromagnetic or radar measurements. One should bear in mind that false depth declarations may cause expensive follow-up activities to fail. This may provoke a general disregard of environmental geophysics. Further mishap may arise through wrong handling of pseudosections. In some cases, the metric scale at the surface (x-axis) was also used to determine the depth of sources. Accordingly, the follow-up drilling was full of surprises.

To avoid this, pseudosections and similar presentations have to be described in detail to the client. A special warning not to measure depths by the meter should be included!

Geophysical anomalies are created not only by natural or artificial structures, but also by human installations like cables or metal pipes, which produce unwanted disturbing anomalies. To avoid this, the area of investigation should be

Table 1.2. Presentation of geophysical results

Medium	Completeness	Readability	Follow-up evaluation possible	Recollection
tape, disc	good	difficult	good	bad
tabled figures	good	difficult	good	small
cross section	acceptable	good	small	good
contour lines	acceptable	good	small	good
3D pictures	small	good	not suitable	excellent
colouring	enhanced	enhanced	not suitable	excellent

checked painstakingly by an electromagnetic cable/metal detector before geoelectric measurements commence.

To eliminate "external" anomalies by calculation is not only very difficult, but often leads to wrong conclusions. If, for instance, a metal pipe crosses under a waste dump, it is preferable to leave the corresponding anomalies in the presentation and show the pipe position in the maps.

An additional obstacle is the meagre knowledge of geophysical methods by environmental experts. This may lead to the wrong approach: "Geophysics must be applied without any background information to find out whether it can really detect the wanted structures." It is much better to disclose all available information to the geophysicist in charge. This will enable him to plan and to evaluate the geophysical work much better and to overcome the ambiguity of some geophysical results.

Most useful is the description of geological, hydrological and tectonical structures and of the materials that have been deposited in a dump.

Geophysics should be employed within a loose frame: the width of the grid and of the measuring array should not be stipulated because all unexpected alterations of form, size and material content of the object, which come up during the survey, must be met with different directions of sections or a change of array.

2 Methods

This chapter describes well-tried procedures that can be employed for the exploration of hazardous waste dumps and their vicinity. In addition, a short description of the principles of measurements, the arrays and the possibilities of application of geophysics to environmental problems are given. Case histories are presented in Chap. 3. More detailed information about geophysical methods can be found in the textbooks cited in Chap. 8.

2.1 Geomagnetic Methods

2.1.1 Geomagnetic Ground Surveys

Magnetic measurements deal with anomalies of the geomagnetic field, which are caused by contrasts of the rock magnetization or by magnetic landfills or dumps.

The magnetization of rocks or iron-bearing waste contains shares of inductive and remnant magnetization: the inductive magnetization originates from the magnetic earth field at the waste location and depends on its actual strength and direction and on the susceptibility χ of rocks or of deposited material. In contrast, the remnant magnetization is constant and is not changed by alterations of the recent magnetic field.

The remnant magnetization is a long-term effect, which is independent of the recent earth field. Only iron and ferrimagnetic minerals can be strongly magnetized. The latter are mostly oxides and sulfides of iron with "spinell structures."

Other materials may be ferro-, para- and diamagnetic. While the similarly strong ferromagnetism is combined with high magnetic susceptibility, the para- and diamagnetism are so weak that they can be ignored in field measurements.

The magnetic effects of magnetic bodies, which can be surveyed on the surface of the earth, are dependent not only on their magnetization, form and size, but also on their depth, because the magnetic field weakens with growing distance from the reciprocal of the power cube.

It follows that the shape of magnetic anomalies flattens with increasing height over the surface of the earth. Figure 2.1 shows this effect for the case of a waste dump with various deposits of different magnetization and size.

The anomalies of singular bodies influence the curves, which are measured close to the surface. Such small-scale anomalies lose their amplitudes rapidly

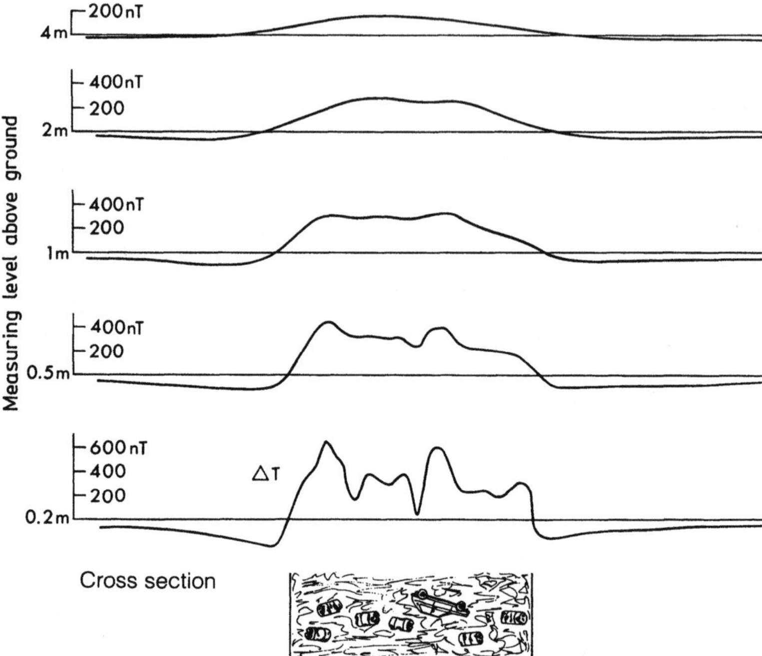

Fig. 2.1. Magnetic anomalies at different heights above ground

with growing height, until finally the curves become so smooth that only the waste deposit as a whole may be discerned.

Therefore, singular iron objects can be located only near the surface. At a depth of > 7 m, even a car wreck can be overlooked.

The shape of magnetic anomalies pertains furthermore to the inclination of the magnetic field of the earth, or the geographical latitude of the location. This is illustrated in Fig. 2.2 for a globe-shaped pile of iron scrap under inductive magnetization.

A typical curve for areas in the northern hemisphere, such as the USA, with inclinations between 60° and 70°, is shown. This example discloses that in such areas, maxima and minima of the magnetic total intensity do not occur over the center of a magnetic source. A normal magnetized body has an anomaly with a strong maximum in the south and a weaker minimum in the north. This must be remembered when follow-up trenching or drilling is planned.

The strength of the magnetic field is measured in nT (nanotessla). Older measurements were recorded in the same numbers, but in gamma. The total field in the USA from the Mexican border to Canada grows from about 43 000 nT to 55 000 nT.

Different types of instruments may be used for magnetic measurements. Some are equipped with permanent magnets, such as the magnetic field balance and the

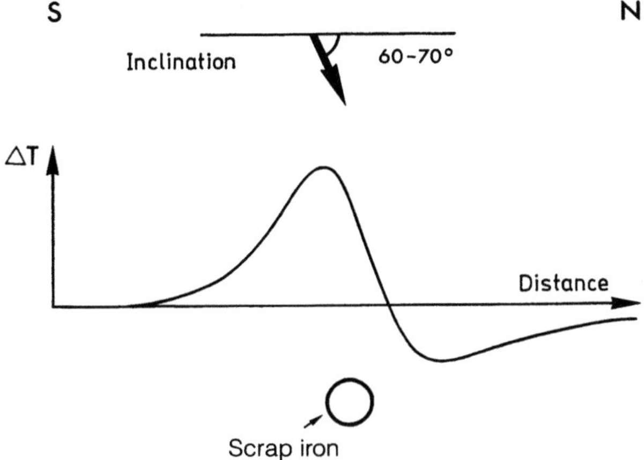

Fig. 2.2. Magnetic section of the total intensity DT over a globe-shaped concentration of scrap iron at 65° latitude

torsion magnetometer. They register the vertical and horizontal components of the magnetic field of the earth. They are of robust construction, but the measurements need a fair amount of time and care. They are used when special information about the shapes and structures of magnetic sources is required.

The Foerster probe relies upon the different magnetization of two ferromagnetic stripes, around which coils with opposing electromagnetic fields are installed. If a magnetic field is applied, the zero point of the two hysteresis curves moves in proportion to the measured field strength. This instrument is used mainly to determine separately the field strength in the horizontal and vertical directions with an accuracy of 1 nT. To achieve this, one Foerster probe is fixed in the vertical and a second probe, in the horizontal position.

Most in use are proton magnetometers, which measure the total intensity T or its variations as ΔT. The principle of measurement is the alteration of the spin frequency of the cores of the atoms of hydrogen, the protons, by the magnetic field. By a strong pulse of approximately 1-second duration, an electromagnetic field is produced in a coil surrounding a tin filled with water, or another suitable organic liquid. This causes the protons of the liquid to spin. After switching off the strong field, the frequency of the spinning protons is registered with an accuracy of 0.5 nT.

Proton magnetometers are simple to handle and allow fast progress of the survey. The measurements of two instruments are often combined to determine the vertical gradient. Two magnetometers are held on the same rod, at different heights above ground (for instance, at 1 m and 2 m). The obtained change of the total intensitiy with increasing height provides the possibility to calculate the depth of magnetic sources with greater exactness.

If a waste dump contains only very little iron or produces only weak magnetic anomalies, the daily variations of the magnetic field have to be considered. They should be continuously monitored at a non-moving base station for which a second proton magnetometer for uninterrupted monitoring of data is needed. The registered variations are eliminated by subtracting the differences in the general total intensity level of the area from the values obtained in the field.

The depth of a magnetic body can be roughly estimated by the half-distance between maximum and minimum of an anomaly. However, it should be observed that the anomalies of only one source are included. For the comprehensive interpretation of magnetic anomalies containing the calculation of form and depth of model bodies, special software is available and should be used.

Geomagnetic data are well geared to localize covered waste dumps. The border of a waste site containing domestic garbage can in most cases be mapped. The same is valid for dumps with a high iron content. Even building refuse sites may contain so much iron that a geomagnetic survey will meet with success.

Furthermore, it is possible to identify particular magnetic deposits in waste dumps, such as bundles of metal drums or scrap iron, on condition that the rest of the waste material is non-magnetic. But the limitations of magnetics are reached when single magnetic objects, like metal drums with toxic fills, have to be located within domestic garbage containing scattered iron objects.

The mesh width of survey grids must be determined by the expected extension of anomalies. Measurements over disposal sites should cover the whole area of rectangular grids meshed 1–5 m. For a preliminary overview, wider meshes can be used.

The magnetic surveys are strongly influenced by artificial magnetic installations like steelmasts, iron posts or steel-enforced concrete, which lie inside or nearby the survey grids.

2.1.2 Aeromagnetic Surveys

Magnetic measurements can also be performed from helicopters or airplanes. Instruments, mostly proton magnetometers or absorption cell magnetometers, are either mounted on the outside of the airplane or are towed behind in an aerodynamic bird. The length of the towing cable varies mostly between 20 and 30 m. Towing has the advantage that the magnetic influence of the airplane must not be compensated.

The flying grid should consist of parallel flight lines in distances of 50–200 m. This line interval should tally approximately with the flying height inside the target area. The points on the ground, to which the magnetic values are to be attributed, should only be spaced 5–20 m. Cross-control lines should be flown in distances 5–10 times the line spacing. At the crossings, the measured data should agree. If this is not the case, the accuracy of the survey may be lacking.

The air survey may be flown in constant topographical height above sea level or at a constant height above ground. The latter is preferred when magnetic bodies near the surface, like hazardous waste sites, are to be explored. The appropriate flying height is 30–50 m. This height is monitored mostly by radar altimeters, which reach an accuracy of 5% of the flying height.

It is necessary to monitor the daily variations of the magnetic field at a base station on the ground during the airborne operation and correct the data, as described for ground surveys.

2.2 Geoelectric Methods

2.2.1 Direct-Current Methods

The DC methods utilize the different electrical resistivities of minerals, rocks and waste deposits. By applying artificial fields of DC (potential fields), the important physical property, the specific electric resistivity ρ, is measured in Ωm.

This method is founded on Ohm's law. It describes the connection between current and voltage if a direct current flows through a conductor of limited size. If a direct current with strength I [A] (amperes) flows along the long axis through a rectangular parallelepiped with cross section q and length b (Fig. 2.3), the voltage between the ends of the parallelepiped is

$$U = I \cdot R.$$

R is called the Ohm resistivity [Ω] and is proportional to the length b and inversely proportional to the cross section q of the conductor and pertains to the specific resistivity [Ωm]. It is valid:

$$R = \frac{b}{q} r.$$

In different arrays of geoelectric surveys, a direct current or an alternating current of low frequency (<100 Hz) is fed into the ground by two metallic current electrodes with low stake resistance. This causes a potential field (Fig. 2.4), which is influenced by the distribution of the specific resisitivities in the earth. By increasing the distance between the electrodes, alterations of the potential field will reflect geoelectric structures at greater depth. Sudden changes of specific resistivity when entering deeper layers of strata will cause characteristic alterations of the curve.

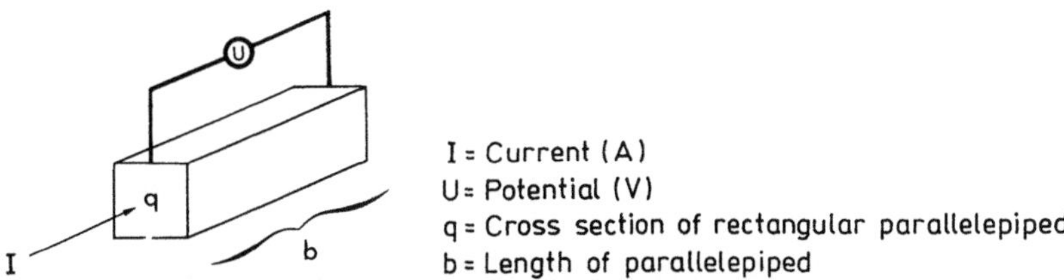

Fig. 2.3. Current flow through a limited conductor

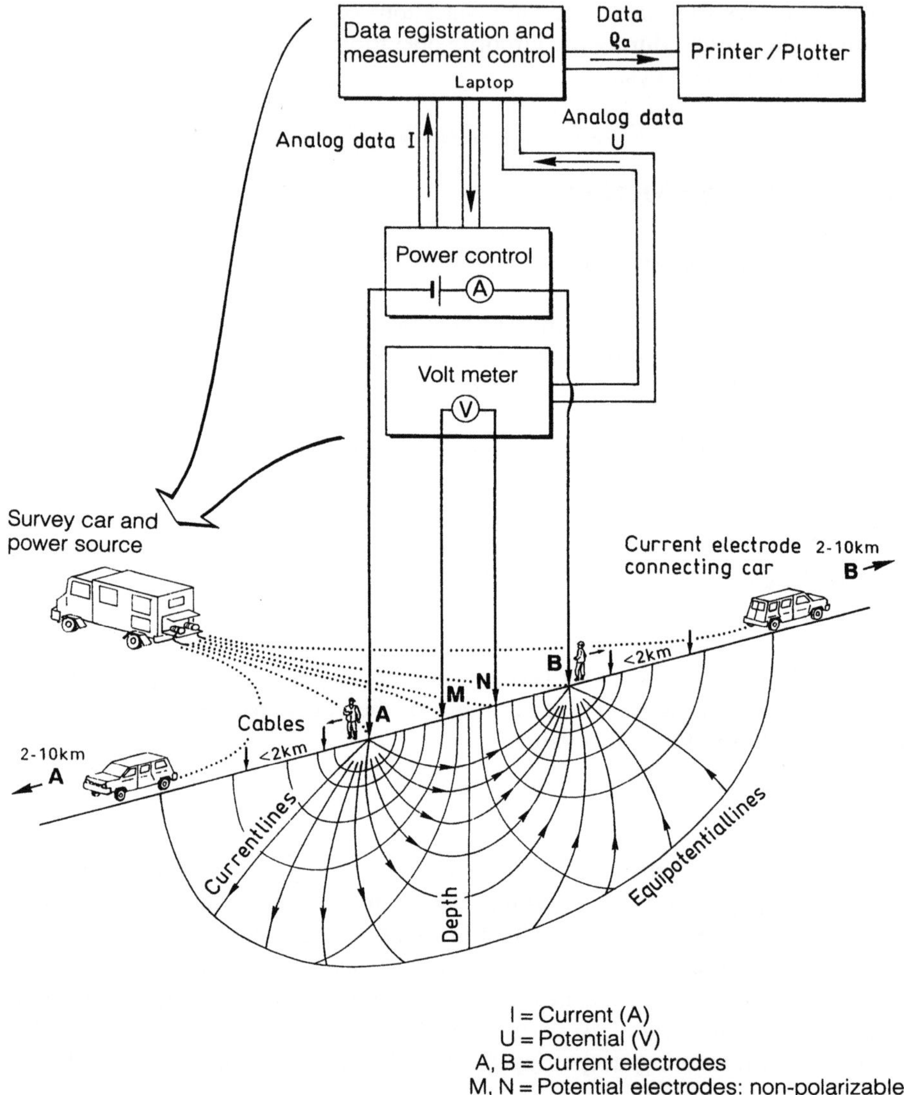

Fig. 2.4. Principle of measurement and potential field for geoelectric DC surveys

Measurements of differences of the potential field (voltage U [V]) are carried out between two well-grounded, non-polarizable potential electrodes. By applying special evaluation software, or by comparing the data with calculated model curves, information about the distribution of specific resistivities and the regarding geological structures can be derived.

Geoelectric field data are normally evaluated and presented by the proportion of voltage U to current I, as measured in the field over inhomogenous ground.

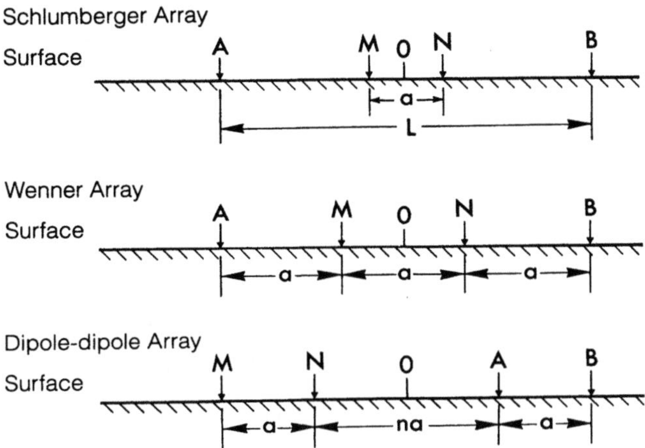

L = AB = Separation current electrodes
a = MN = Separation potential electrodes
0 = Point of measurement

Fig. 2.5. Arrays for geoelectric mapping and sounding

These data are converted into values, which would be valid for a homogenous half-space by considering the actual electrode array. These values are called "apparent" specific resistivities ρa and are expressed in [Ωm].

This transformation is done by multiplying the specific resistivity ρa with the geometric "K factor". Prominent K factors for customary geoelectric arrays (Fig. 2.5) are:

$$\rho_s = K \cdot U/I \ [\Omega m]$$

$$K \text{ Schlumberger} = \pi/a \left[(L/2)^2 - (a/2)^2 \right]$$

$$K \text{ Wenner} = 2\pi a$$

$$K \text{ dipole-dipole} = \pi a \cdot n(n+1)(n+2)a$$

L = spacing of current electrodes
a = spacing of potential electrodes
n = multiple of a

This list of specific resistivities is based on various singular investigations in Europe and the U.S.A. It is to be regarded as a rough guide only, since deposited material may, for instance, be mixed, which may result in very different resistivities. On the other hand, it shows how wide the scope of geoelectric surveys can be.

Geoelectric DC methods are predominantly used for:

– *Geoelectric mapping:* determination of the horizontal distribution in defined depth horizons.
– *Geoelectric sounding* (also electrical or resisistivity sounding ES): investigation of specific resistivity and thickness of horizontally layered strata.

Table 2.1. Specific resistivities

Rock type/Material	Specific resistivity [Ωm]
Rock type	
clay, marl, rich	3 – 30
clay, marl, meagre	10 – 40
clay, sandy, silt	25 – 150
sand, with clay	50 – 300
sand, gravel in ground water	200 – 400
sand, gravel, dry	800 – 5000
rubble, dry	1000 – 3000
limestone, gypsum	500 – 3500
sandstone	300 – 3000
salt beds and salt domes	> 10000
granite	2000 – 10000
gneis	400 – 6000
Deposited refuse	
domestic garbage	12 – 30
debris and dumped soil	200 – 350
industrial mud	40 – 200
scrap metal	1 – 12
pieces of broken glass and porcelain	100 – 550
casting sand	400 – 1600
wastepaper (wet)	70 – 180
contaminated plume of domestic-garbage dump	1 – 10
used oil	150 – 700
tar	300 – 1200
cleaning clothes and materials	30 – 200
used lacquer and paint	200 – 1000
barrels (empty)	5 – 20

Geoelectric Mapping

Lateral differences in apparent specific resistivity are mapped for a distinct depth
level. Suitable targets are, for instance, the rims of disposal sites or of deposits of
contaminated materials. The data are obtained from a fixed array, which records
the potential differences between the potential electrodes and is moved step-by-
step along survey lines until the whole survey area is covered. The result is presented as a contour map or resistivity section.

The depth range of the survey should not extend beyond the body of the waste
deposit into the bottom rock. For this purpose, the Wenner array (Fig. 2.5.) is well
geared. It can detect the border of most domestic waste sites very accurately.
However, the condition has to be fulfilled that the low specific resistivities that
are common to domestic refuse do not agree with similar low resistivities of the
surrounding country rock.

This condition is met by rocks with apparent specific resistivities from 300
to > 2000 Ωm, which are much higher than the resistivities of waste (< 20 Ωm).

Gravel, sand, limestone and sandstone are well suited, but clay and marl may possess apparent specific resistivities of the same order. Waste deposits lying within such low resistive rocks may therefore remain undetected.

Apart from this special case, even separate contaminations with deviating resistivities, like casting sands with very high, or galvanic muds with very low resisitivities, can be located by geoelectric mapping. The resistivity values in Table 2.1 may be used as a guide to materials that can be found by geoelectric mapping.

The chosen array influences not only the result but also the length of survey time. The most efficient arrays for mapping are presented in Fig. 2.5.

Important obstacles for geoelectric mapping are metallic cables, pipes and other installations. They must be located by special metal- or cable-detectors before the survey. If this is not done, their very strong artificial anomalies may delude the geophysicist into making an erroneous evaluation.

It is certainly advantageous to carry out geoelectric mapping for more than one depth. This is done by increasing the distance between the electrodes. However, the following rules of thumb should be observed:

- The spacing of current electrodes should agree with the double depth of penetration.
- The spacing of potential electrodes should tally with the desired depth penetration.

Geoelectric mapping ought not to be confined to waste sites, but may also be useful in their vicinity. The lateral extension of aquifers in unconsolidated rocks, like water-saturated sands and gravel with high specific resisitivity, can be found against low-resistive clays and marls. The spread of salty leachate plumes can also be monitored.

Geolectric Sounding (VES)

This method is used to determine:

1. The apparent specific resistivities of horizontally bedded strata or deposited material.
2. The thickness and/or depth of those boundaries where resistivities of beds or dumped charges change.

In use are 4-electrode arrays that are based on an artificially produced stationary electric field. An often applied electrode configuration is the "Schlumberger array" (Fig. 2.5). But the Wenner and dipole-dipole arrays may also be used for geoelectric sounding.

Two well-grounded current electrodes feed a DC current into the earth with the current of I. The potential differences U are measured between two neighbouring potential and non-polarizable electrodes in the center of the array. To successfully gather information about the depth of the underlying beds, many (>20) measurements are necessary, with distances between the current electrodes increased according to an approximately logarithmic scale.

From the obtained data of the current I, the voltage U and the geometric factor K, the apparent specific resistivity ρa is calculated. All ρa-values are plotted on log-log graph paper against the half-distance of the current electrodes (L/2). The single points are then connected to a sounding curve (Fig. 2.6).

The evaluation of the sounding curve can be obtained in several ways:

1. by the manual auxiliary-point method for multilayered cases utilizing special curve sets,
2. by comparison with sets of printed master curves,
3. by using digital software with inversion programs.

The result should always be the determination of the following:

1. the number of beds,
2. the thickness of individual beds,
3. the resistivity of individual beds.

Geoelectric sounding has two important limitations:

1. omittance of beds:
very thin beds cannot be derived from the sounding curves at greater depth.

2. the principle of equivalence:
An evaluation of a sounding curve may produce several equivalent solutions. The geophysicist has to select the result that agrees best with the known geological and hydrological structures of the ground. Another selective moment is the comparison with neighboring soundings. It must be possible to connect the depth marks of the boundaries of single layers of one sounding to corresponding depth marks of the next sounding in such a way that a geologically or environmentally plausible section is created (Fig. 3.13).

Sounding curves of > 2 layers are subdivided into four types:

1. minimum (H),
2. maximum (K),
3. double-descending (Q),
4. double-ascending (A).

Graphs pertaining to the auxiliary-point method are available. When evaluating ascending curves (nos. 2 and 4), anisotropy effects have to be considered.

When constructing geoelectric sections or isoline maps of the boundaries of single beds, correlations must be limited to horizontal or flat, dipping strata. Steep dipping horizons or fault planes cannot be constructed by one-dimensional evaluation. They must be either drawn by using geological knowledge or by applying computer programs for two- and three-dimensional interpretation.

If a surveyed area does not possess a homogeneous, horizontally layered ground, disturbing side effects may derange the evaluation considerably. It was a surprise that the very inhomogenous dumps of domestic waste show unexpectedly uniform low resistivities (Table 2.1). Therefore, it is feasible to determine pre-

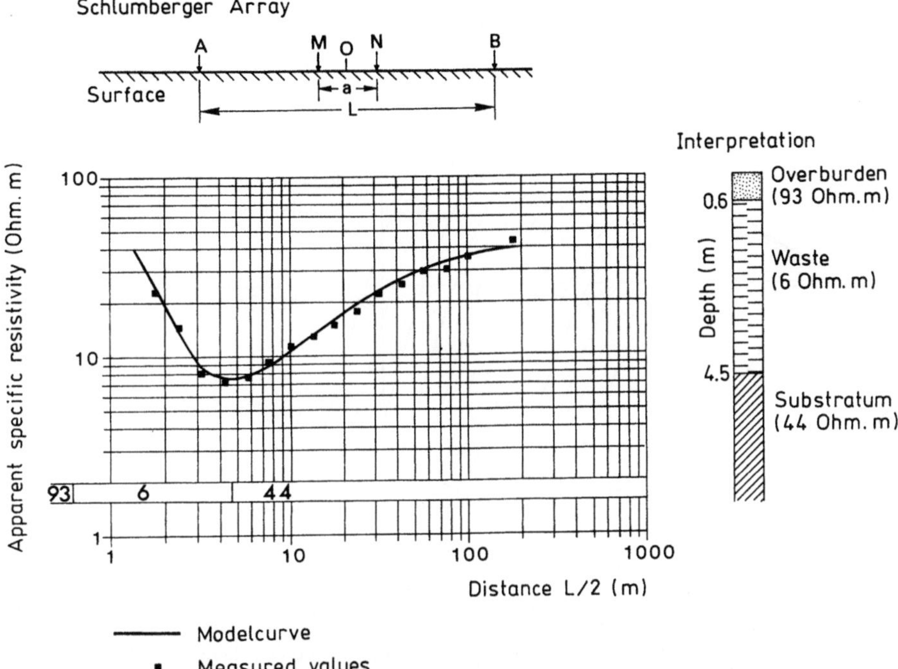

Fig. 2.6. Geoelectric sounding curve (VES) of a Schlumberger array with digital interpretation and computed model curve of the minimum type "H"

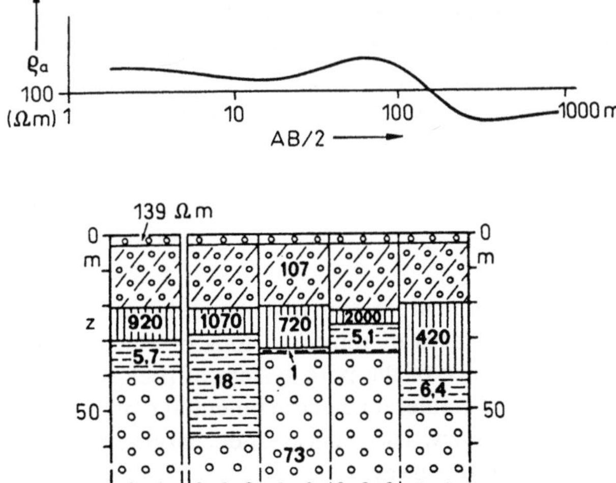

Fig. 2.7. Equivalent digital interpretations of a Schlumberger sounding curve. Left column = mathematically best model. The selection of the most suitable model has to consider neighboring curves and the known geology

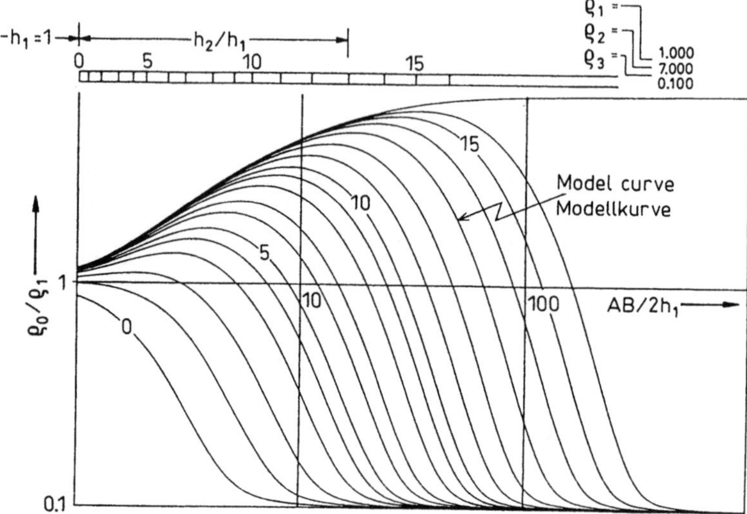

Fig. 2.8. Three-layer master curves in a log-log graph of the INGESO atlas. The resistivities of the three beds are in the ratios 1:7:0.1; first layer : second layer : third layer. The sounding curve, which has been drawn on log-log graph paper in the field, is laid on top of the master curve and moved around until one of the master curves tallies with the field curve. The thickness of the second layer, which has here seven times the ρa-value of the first layer (see the resistivity values at the top right) can be found by the number of the curve no.13. On the thickness beam at the top left, which is divided from 0 to 16, the thickness h_2 can be directly determined

cisely the depth, thickness and specific resistivities of a domestic disposal site by geoelectric sounding.

Since ammonia and other gases of decay are emanated by domestic waste, work safety is impaired if invasive methods of investigation, like drilling or probing, are applied, because the inevitable destruction of the top sealing often results in outbursts of gases that might endanger the health and lives of the workers.

As a non-invasive method, geophysics is therefore certainly less dangerous and much safer to apply.

Induced Polarization

The IP method is based on an electrochemical reaction of a natural liquid solution (for instance, ground water) with interfaces of minerals. Ions or electrons are caught on the walls or the interior surface of the pore system of rock, when an electric DC or AC field is applied. This creates an electric potential that could be compared with the charging of a car battery. The process of decharging provides the parameters, which are measured by IP surveys.

Two types of induced polarization are known:

1. Metallic Polarization

This occurs at the surface of minerals with high conductivity and metallic luster.

2. *Boundary-Layer Polarization*

It is much weaker than the metallic polarization and develops an accumulation of cations at the boundary layer of the electrolyte, which spreads over the pore walls of silicate rocks.

For environmental application, it is important that saline waters prevent induced polarization, since their high conductivity does not allow for any ion accumulation. This effect permits the discrimination between saline ground water and clay with the same specific resistivity: IP signals disappear when they enter saline water, but may grow or remain steady when going through clay beds.

The arrays of IP surveys are basically the same as for DC methods. Through two current electrodes, a direct current is fed into the earth. This causes the induced polarization or the charging of the subsurface. When the current is cut off, the voltage at the potential electrodes does not drop immediately to zero, but reaches a secondary lower level within 30 ms. From there, the measurement of the transient voltage starts.

The curve of this transient decay is recorded by the IP receiver at several time intervals. The measured parameter is called the chargeability M. It is defined as the average of a specified time interval of the decay curve. Mostly, ten intervals are registered at logarithmic intervals. The length of the measurements at the transient decay curve can vary from 1 to 8 s. Most common are the total decay times of 2 and 4 s. At the end of the decay time, the current is switched on but reversed to erase remanent charges, and the same procedure is repeated in the opposite current direction. Figure 2.9 portrays this process.

Chargeability values are dependent not only on the distribution of chargeable bodies in the earth, but also on the strength of the primary current. It is therefore not possible to compare chargeabilities without knowing the primary current and the length of the recorded time intervals.

The first section of the decay curve, up to 10 s after the switching off of the primary current, is dominated by electrodynamic processes. During this first period, charging of the earth is weaker than the induction processes that disturb the IP signal. Therefore, the registration of the transient timeintervals should start only after this inductive period.

Every IP measurement includes not only the gauging of the chargeability, but also the determination of the primary voltage V_P and of the self potential Sp. V_p is used in the same manner as in the geoelectric DC methods to calculate the apparent specific resisitivity ra (see Sect. 2.2.1). For this, no extra costs and no additional time is necessary.

When planning an IP survey, it is important to know that the decay voltage reaches generally only 1–2% of the primary voltage. Since in many areas industrial and/or telluric noise is very strong, the primary current should be increased to keep the transient voltages above 1 mV. Otherwise, the recorded decay-curves may reflect more noise than effects from underground.

All possible arrays known from DC geoelectrics can be used for IP surveys, but the most employed IP array is dipole-dipole. Its advantage is its symmetry: the IP

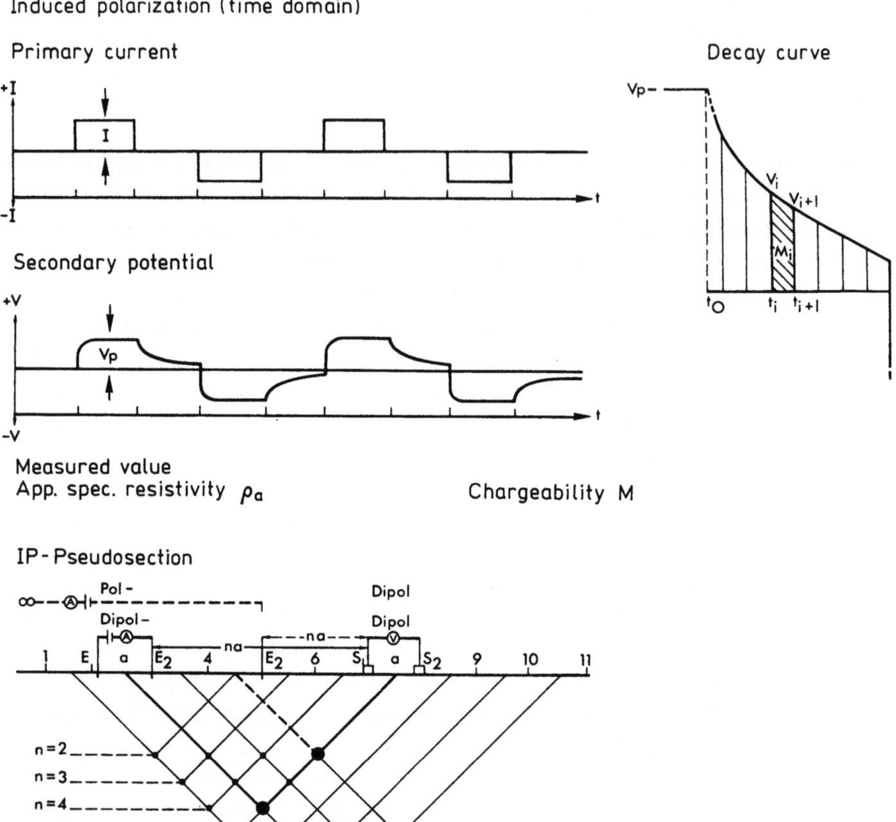

Fig. 2.9. Principle of induced polarization (IP)

sources lie under the center of the array and clues as to their depth and dip may be derived. Additionally, better safety is reached, since a short dipole of dangerous current electrodes can be more easily controlled than any current electrode at great distance.

The depth penetration of IP surveys depends on the length of dipoles a and the distance $n \cdot a$ of the two dipoles (Fig. 2.9). The length a should be at least twice and at most ten times the thickness of the expected IP body to achieve a proper result. The geometric depth-extension is calculated by

$$d \sim a \cdot (n+1)/2.$$

Customary dipole lengths are $n = 1$ to $n = 6$. Further enlargements may lead to an unacceptable signal noise relation, which may torpedo all IP measurements.

The evaluation starts with the conversion of IP field data into the apparent specific resistivity, the chargeability and perhaps the self-potential. The latter is often omitted because changing of potential electrodes after "salting" the current electrodes may result in erroneous observations. The next step is the interpretation of the evaluated data. It should result in estimations of depth and shape of the surveyed body. From the chargeability values, decay coefficients and potential functions may be derived, which makes it possible to draw conclusions about the IP source.

Environmental applications of IP surveys are the determinations of the borders between hazardous waste and the surrounding uncontaminated soil, the locating of special deposits of chargeable material in a dump, and the mapping of contaminated salty plumes in clayish country rock.

However, knowledge of specific chargeabilities of hazardous and toxic materials is still lacking. Increased chargeabilities were hitherto observed in galvanic mud, printed paper, glazed pieces of pottery, rusty scrap iron , non-iron metal remnants, buried shells and grenades, casting sand and electronic refuse.

Self-Potential

Natural geoelectric fields are called self-potentials. They, or rather their distribution on the surface of the earth, may provide information about inhomogenities underground. Self-potentials originate from electrochemical processes that occur when ores or buried metals contact rocks, ground water or rock fluids. Such reductions and oxidations are based on static contacts and are called "redox potentials". They are normally > 30 mV and may reach > 200 mV.

Other self-potentials emerge from the rapid movement of water or gases in soil or rock. They are called flow potentials and arise by fast ground water flow and strong infiltration of waste-dump seepage. Flow potentials are mostly weak (< 10 mV) and therefore difficult to distinguish from industrial and telluric noise. Figure 2.10 shows the two potentials that may be found in a hazardous waste dump.

Self-potential surveys may be used with some reservations for the detection of oxidizing metals in dumps or soil. However, the overlapping redox and flow potentials cause an element of uncertainty in interpretation that permits only qualitative statements.

Nevertheless, the locating of gas eruptions through sealings by Sp measurements has proved to be successful. The observed strong flow potentials vary considerably with time and can thus be discriminated from redox potentials, provided a dense grid is laid and is often surveyed in minute intervals.

Daily variations of the natural potential field are overlaid by longer periodic changes of telluric origin. Other disturbing influences are man-made. They often come from electrified railroads, where locally short pulses (> 30 mV) were measured. To eliminate all of this, the continuous monitoring of the electric potential at a base station is compulsary because the raw field data must be corrected. This is especially necessary if a great number of non-polarizable potential electrodes are registered simultaneously.

An advantage of Sp surveys is their simplicity: between two potential electrodes, only the voltage must be registered accurately. One potential electrode has to be the base for all measurements and should be grounded in an electrically un-

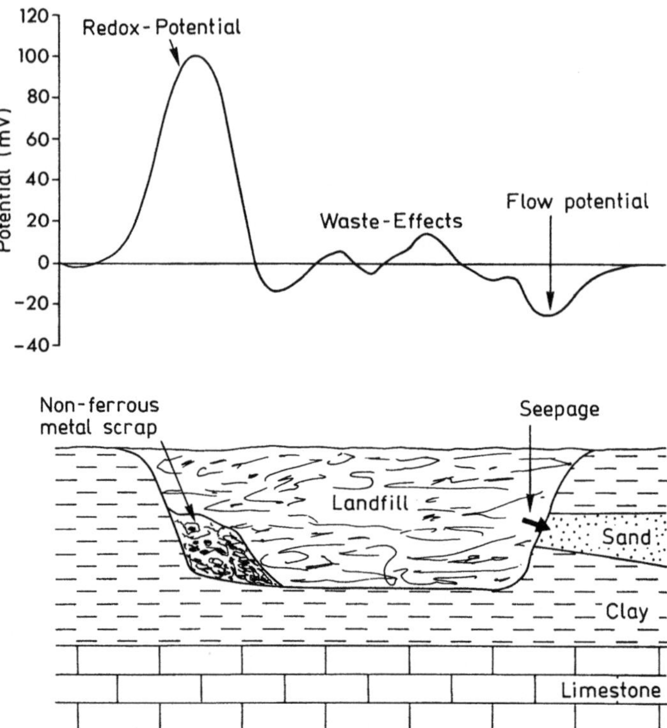

Fig. 2.10. Pattern of different Sp anomalies over a hazardous waste site

disturbed area. The second electrode could be moved along the survey lines, or many secondary electrodes could be recorded at once when surveying by the scanner method.

2.2.2 Electromagnetic Methods

Electromagnetic Mapping (EM)

Mostly in use are EM surveys with movable, horizontal and coplanar coils. Vertically polarized sinus oscillations (also known as "Slingram") are transmitted and received. An alternating primary field is transmitted, which induces in rocks with dissimilar specific electric resistivities eddy currents or secondary fields. These can be directed either against or parallel to the primary field, depending on whether the resistivity of the body is lower or higher than the surrounding rock or material. By interference of the primary and secondary field, the resulting field is produced, which is finally received on the surface.

The principle of electromagnetic surveying is presented in Fig. 2.11. Changes of the in- and outphase components allow conclusions as to the positions of especially good or bad conducting bodies underground. The inphase component (or

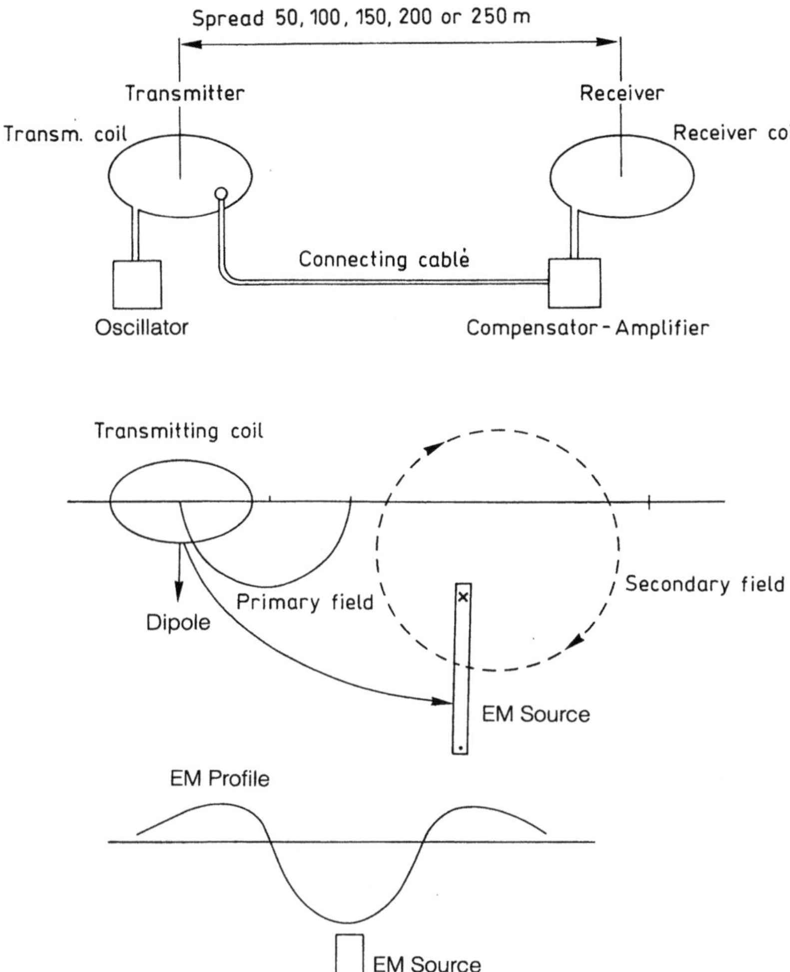

Fig. 2.11. Principle of electromagnetic mapping

real phase component) is this portion of the alternating field that oscillates in the same phase as the primary signal. The part of the primary field whose phase is rotated by 90° is called the outphase component (or imaginary component). Determination of the depth of the investigated body can be based on the measurement of many frequencies by the use of the "skin effect," as shown in the nomogram in Fig. 2.12.

Some instruments can transmit up to 12 frequencies. The receiver amplifies the received resulting field and compares it with the primary field, which is obtained from the transmitter by a connecting cable. Even small changes in the distance between the receiver and the transmitter may influence the values of the inphase

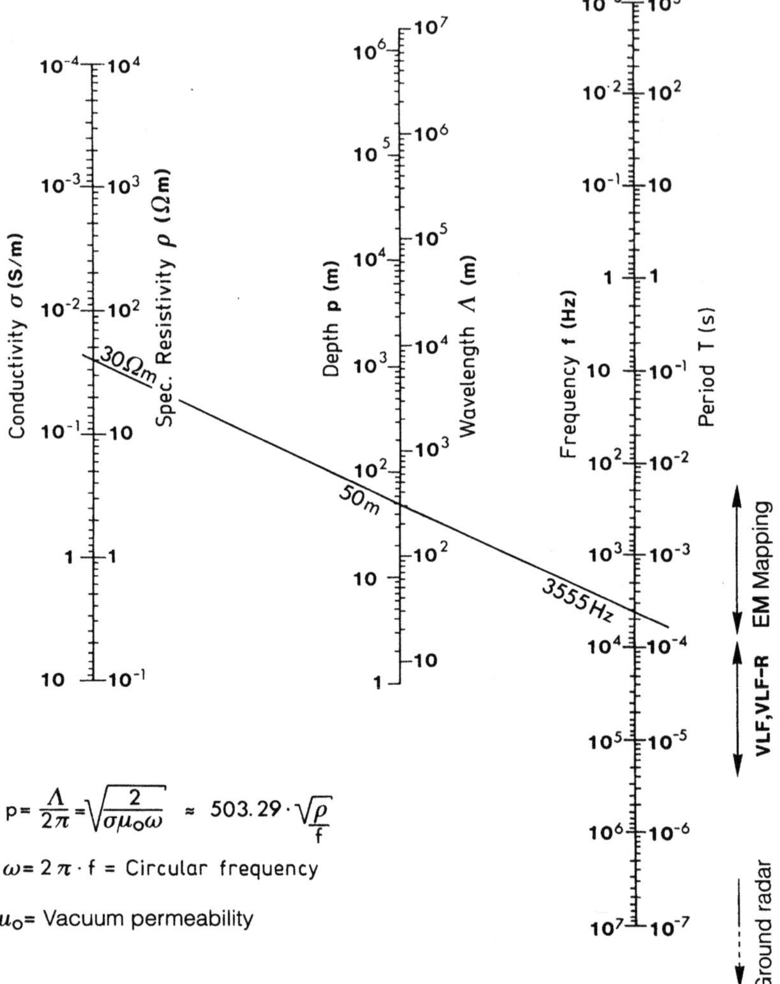

$$p = \frac{\Lambda}{2\pi} = \sqrt{\frac{2}{\sigma\mu_o\omega}} \approx 503.29 \cdot \sqrt{\frac{\rho}{f}}$$

$\omega = 2\pi \cdot f$ = Circular frequency

μ_o = Vacuum permeability

Fi. 2.12. Nomogram showing the relations of specific resistivity (left column), depth of penetration (middle column) and frequency (right column) of a homogenous plane wave

component. Hence, all measurements in mountainous terrain must be slope-corrected.

The point to which the surveyed data belong always lies in the middle of the distance between the transmitter and the receiver. The separation of single survey points should not exceed one-quarter of this distance. The positioning of the survey can be done by compass and chain; directions and distances to landmarks like crossings, railways, roads, buildings and trigonometric points must be entered into field maps with such accuracy that follow-up drills can be positioned in an orderly way. In areas with no or faulty maps, the satellite-positioning system GPS should be employed.

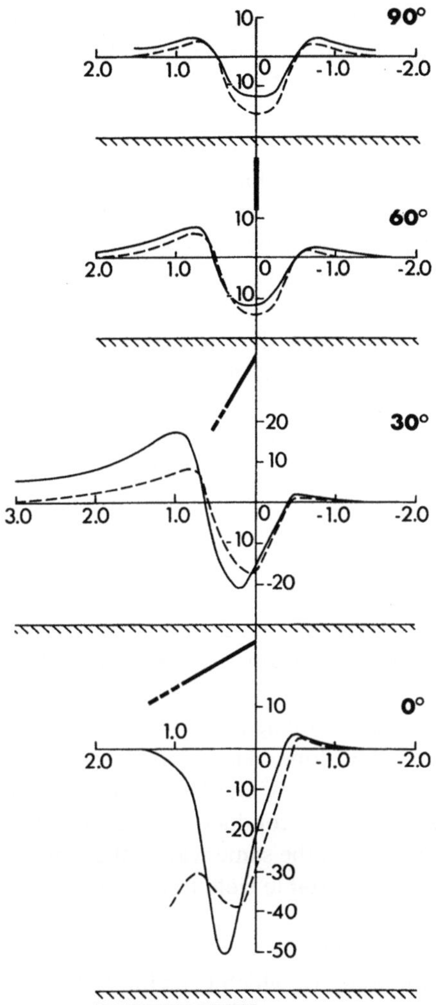

Fig. 2.13. Typical EM curves over steep dipping, conducting sheets like faults or fracture zones with different dip

Buried metallic cables or pipes create electromagnetic anomalies that disturb any EM survey. To avoid this, the survey area has to be closely checked by a cable or pipe detector before the measurements begin. If disturbing bodies are found, the survey lines should be relocated with a minimum distance of one half the transmitter-receiver distance from the disturbance.

The relation between the specific resistivity of rock, the depth penetration and the frequency of a flat homogenous wave is portrayed by the nomogram in Fig. 2.12. The line from 30 Wm to 3555 Hz, for example, crosses, the depth of 50 m. If the frequency increases to 10000 Hz, the wave would propagate to a depth of only 20 m. The conclusion to lower the frequency in order to reach greater depths is correct but to transmit lower frequencies, more energy is needed. For movable instruments, this is not possible. Therefore, most instruments do not go below 100 Hz. The frequencies between 800 and 7000 Hz are the best to reach appropriate depth penetration under field conditions.

EM surveys are especially adapted to mapping lateral differences of resistivity or conductivity. Possible targets are the borders of waste dumps, singular objects and steep dipping structures, such as faults, fracture zones or crevices, which may guide contaminated leachates. By correlating EM anomalies from profile to profile, "linears" of such zones of tectonic weakness can be constructed. But the known tectonic pattern of the area has to be taken into account.

Since light-weight EM instruments have to be carried during survey, they cannot be powered by strong and heavy batteries. The already mentioned limitation of depth penetration results also from this fact.

Figure 2.13 shows typical electromagnetic models of the in- and outphase components over steep dipping, good conducting sheets like faults or fracture zones. The dip of these bodies obviously influences the shape of the EM curves.

Electromagnetics by Distant Transmitters (VLF)

A special EM application is the Very Low Frequency (VLF) method. It uses frequencies between 12 and 25 kHz, which are very low from the point of view of electromagnetic communication technology. For geoelectric use however, they are very high because they do not penetrate deep into the earth (Fig. 2.12).

The VLF sinus waves are sent from several permanent, very strong transmitting stations, which are scattered around the globe. Their mostly constant signals are used mainly for the navigation of submarines. But besides this original purpose, they induce secondary fields in electric conducting bodies in the ground. These homogeneous fields can be utilized in the same way as the heterogeneous EM fields, which are produced by small portable field transmitters. The following table lists some VLF stations:

In spite of the fact that VLF saves the application of a special transmitter, the method has some disadvantages, the most important of which is its limited depth penetration. In clay-bearing rocks with resistivities of < 30 Wm, only a depth of

Table 2.2. VLF Stations

Station	Location	Frequency (kHz)	Output (kW)
FUO	Bordeaux, France	15,1	500
GBR	Rugby, Great Britain	16,0	750
UMS	Moscow, GUS	17,1	1000
NAA	Cutler, Maine, USA	17,8	1000
NLK	Seattle, Washington, USA	18,6	300
IVC	Tavolara, Italy	20,3	500
NWC	NW Cape Australia	22,3	1000

approximately 15 m can be reached. Apart from that, the nearly homogeneous field often omits small random structures, which are common EM targets in hazardous dump sites, however.

The survey lines should be laid at a right angle to the straight line between a VLF transmitter and a survey area to achieve high anomalies and to avoid distortions (Figs. 2.14 a and b). When the described restrictions are followed, the VLF method can be exercised in the same way as EM surveys with heterogeneous fields.

The VLF-R method, also called VLF resistivity or radio ohm method, gauges the resistivities of the ground. In addition to the horizontal alternating electric field on the surface, the perpendicular horizontal component of the magnetic field in the air is measured. The following parameters can be derived:

1. real part of the vertical and horizontal components of the magnetic field, arising from currents that are induced in conducting bodies;
2. imaginary part of the vertical and horizontal components of the magnetic field, arising from currents that are induced in conducting bodies;
3. magnetic total intensity;
4. apparent specific resistivity;
5. phase differences between horizontal electric and magnetic field components.

Such large topographical effects can be avoided in flat areas. Otherwise, those effects have to be negated by special software to produce dependable and comparable results. In spite of this detrimental dependence on the direction of the VLF station, one should not overlook the advantages of the method: it is rather fast and very cheap; furthermore, it provides the possibility, when using many VLF stations, plus long wave radio stations up to 240 kHz, of performing EM multi frequency depth soundings.

The raw field data have to be corrected for diurnal and short periodic variations of the intensitiy of the VLF field. It is therefore essential to monitor the alternating electromagnetic field during the survey time. By doing so, all off-times of VLF stations are recorded.

VLF is well geared to detect buried hazardous waste, even under extended areas, quickly and accurately. The time needed for one measurement is only 1 m. New instrument-combinations with proton magnetometers will make it possible to kill two birds with one stone.

Magnetotelluric and Transient Electromagnetic Methods (MT + TDEM, TEM)

The *audio magnetotelluric method* (AMT) measures frequencies between 1 and 20 000 Hz, which are abundant in the ground, stemming from lightning or other natural sources. The ratio of the horizontal electric field to the magnetic field is measured and calculated as a function of the applied frequency. The result is the apparent resisitivity depending on the frequency. This method provides better results if an artificial field source is applied by a remote grounded bipole transmitter. This is called the *controlled-source audio magnetotelluric method* (CSAMT). Both systems are rarely used in environmental explorations.

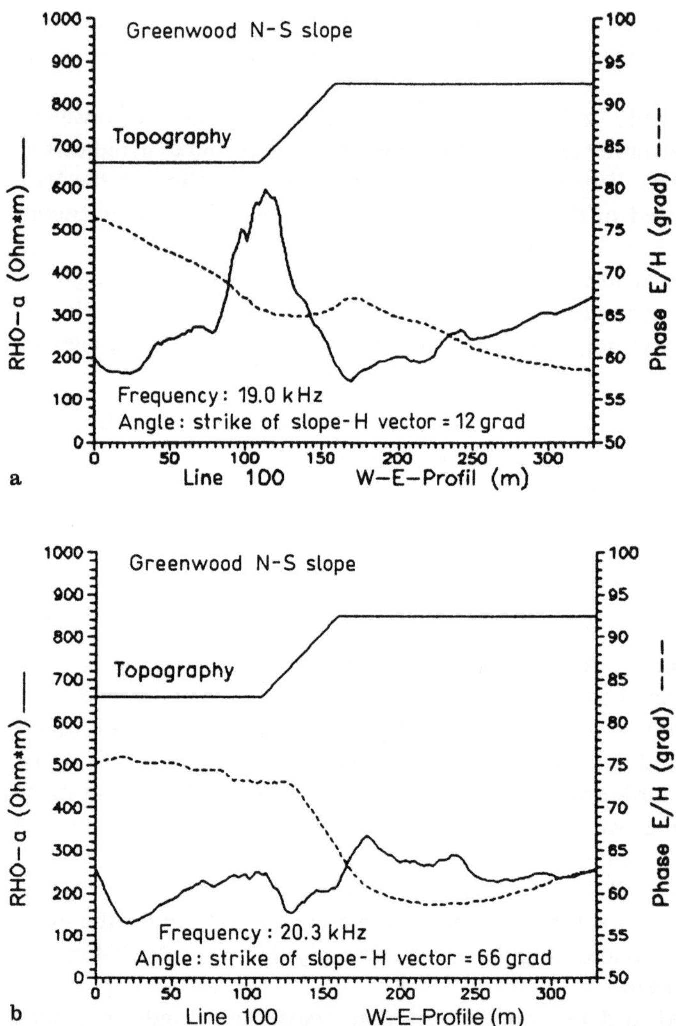

Fig. 2.14a. Influence of topography on VLF surveys. Pictured are the apparent resistivity rs and the phase difference E/H between horizontal electric and magnetic fields. The VLF station transmits a frequency of 19.0 Hz out of an angle of 12° to the strike of the slope, which runs perpendicular to the plane of the book [2]

Fig. 2.14b. Changing of VLF stations alters the topographical effects so much that results cannot be compared among stations. The same terrain as in Fig. 2.14a was surveyed by a VLF station with a frequency of 20.3 Hz and an angle of 66° to the strike of the slope

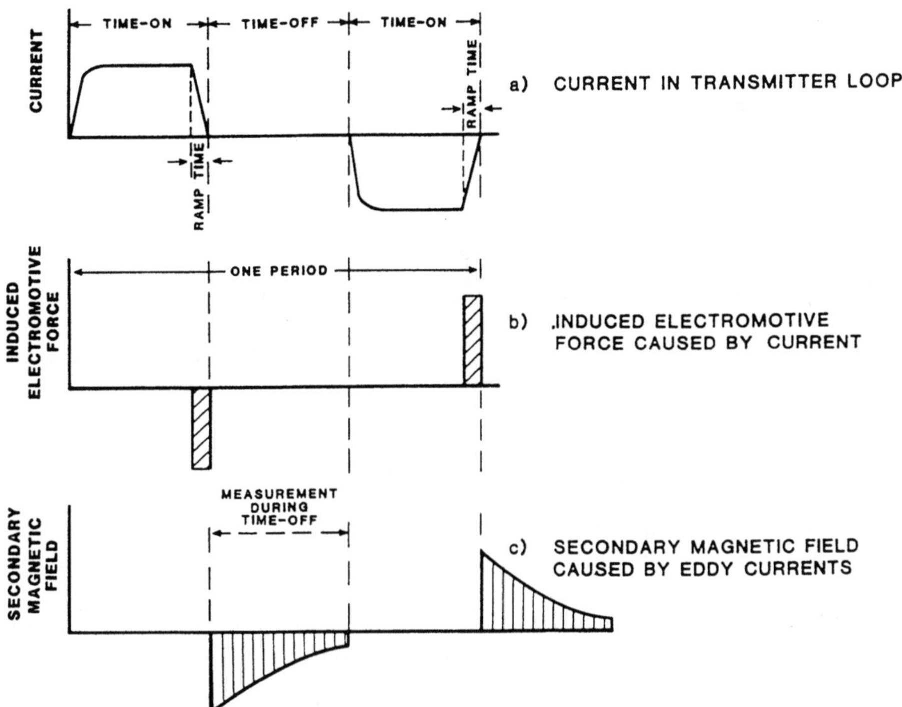

Fig. 2.15. System wave forms employed by the TDEM method

The *time domain* or *transient electromagnetic system* (TDEM or TEM) must not be confused with the time domain IP. TDEM measurements are also based on the transient decay of a current with time, but here the anomalous magnetic field (not the transient voltage) is recorded in a very short period. The observations commence only microseconds after the shut-off of the primary current, which is induced into the ground by a transmitting loop of 5 to >100 m diameter.

TDEM soundings have a wide scope from <50 m to >1000 m depth. Their results are equivalent to VES (DC resistivity sounding); an advantage is the short length of the total array and the very great depth penetration with a small loop size. They have been successfully applied to map the extent of brine plumes and to determine the encroachment of salt water into fresh water aquifers.

Airborne Electromagnetics

The sensors for electromagnetic measurements by airplane or helicopter are mostly towed by a "bird", an aerodynamically shaped container, which is suspended from an aircraft by a $\sim$30 m cable. The bird is 6–10 m long and encloses a transmitter and a receiver coil at both ends. The axes of the two coils stand perpendicular to each other. The principle of this method is the same as for EM ground surveys.

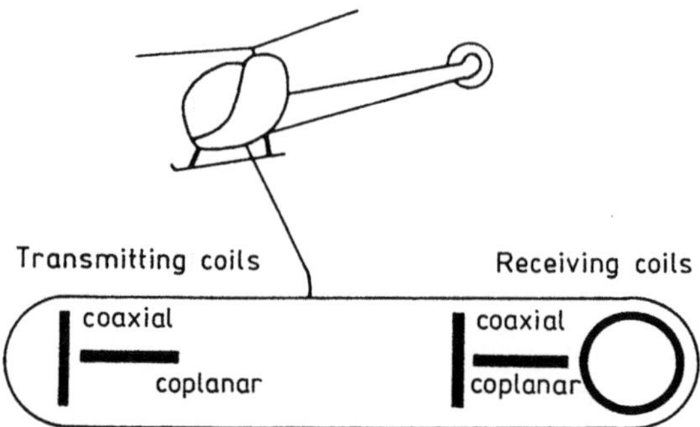

Fig. 2.16. Diagram of an EM helicopter survey (Dighem system) with maximum coupling and a coil separation of 8 m

Because of the small transmitter-receiver separation and the large distance between the transmitter in the air and the surveyed object on the ground, the measured values of the residual field are very weak. Therefore, a much higher degree of accuracy is demanded; the precision of measurements has to be better than the cube of 3.

A grid of parallel lines is flown, whose distances have to match the size of expected EM sources. Customary are intervals of 30 to 50 m. The airborne results are presented in contour maps of apparent specific resistivities. This parameter is derived from the registered in- and outphase values by special evaluation software. In addition, statements concerning the shape and depth of conductors can be made by special model calculations.

Good results of airborne EM were obtained by the mapping of the freshwater interface in arid areas and near coastlines. In general airborne EM surveys, good conductors may be located at depths of up to 200 m, provided the overburden is of high resistivity. When planning airborne EM for environmental purposes, its high cost has to be taken into account. Only large areas should be flown under one contract to achieve an economical return. For investigations of hazardous waste sites, airborne methods are exceptional. If employed, airborne EM must be combined with aeromagnetic and aeroradiometric surveys (Sect. 2.1.2).

Georadar

Georadar is also known as the ElectroMagnetic Reflection method (EMR). It is employed as ground radar to investigate structures at very shallow depths. In salt domes, permafrost areas and cristalline rocks of extremely high resistivities, seepages or concentrations of lye are located by this method. It is based on the reflexion of high-frequency electromagnetic waves, from 8 MHz (megahertz to 4 GHz (gigahertz), at interfaces of materials where the dielectric constant ε and the conductivity change.

ε is a measure of the capacity of a material to store charges when an EM field is applied. It is the dimensionless ratio of the capacitance of the material to that of free space. Examples of dielectric constants are listed in Table 2.3.

Table 2.3. Dielectric constants (K), electric conductivity (σ), electric velocity and attenuation (a) at a frequency of 100 MHz. Davis and Anan (1989)

Material	K	σ(mS/m)	v (m/ns)	a (dB/m)
Air	1	0	0,3	0
Freshwater	80	0,01	0,33	$2 \cdot 10^{-1}$
Seawater	80	$3.0 \cdot 10^4$	0,01	0,1
Dry sand	4	0,01	0,15	0,01
Wet sand, Aquifer	25	$0,1^{-1}$	0,06	0,03
Limestone	6	$0,5^{-2}$	0,12	0,04
Fat clay	5–35	0,05	0,06	1,0–300
Granite	5	0,1–1	0,13	0,01
Rock salt	6	0,1–1	0,13	0,01
Slate	5–15	0,03	0,09	1,0–100

The depth of penetration is limited by low conductivity (or by its reciprocal high resistivity) of the ground. The signal is for instance attenuated by conducting clays to a depth of only 0.2 m. But in salt, ice or dry granite, > 300 m may be penetrated.

Since water has a high dielectric constant of 80, changing moisture of soils and rocks influences the radar response considerably. The same is valid for varying conductivities by alternating clay contents of soil, which alters the depth of reflections. Therefore, radar measurements before and after rainfall may produce very different results. Interpretations of ground radar should therefore be made with these restrictions in mind and should, if possible, be verified by another method.

The radar signals are transmitted as pulses of the high frequencies. The receiver antenna registers those pulses that were not absorbed by the ground but were reflected in exact time dependence. In ground radar surveys, transmitter and receiver are coupled and are pulled together across the terrain on a sledge. This leads to the registration of continuous radar sections. The principle of measurement is shown in Fig. 2.17.

The signals, which are reflected by horizons of discontinuity, are recorded after a certain travel time, which depends on the surveyed material. This method is not only similar to seismic reflection, but it is even possible to use the software for seismic evaluation to interpret ground radar measurements. As in seismic work, the depth of the reflector can be concluded if the velocity is known.

Radar measurements are a fast method to detect small objects close to the surface of the earth (0.1 to ~ 3 m) with high resolution. Surveys should be made only over a dry and nearly homogeneous ground with high electric resistivity, low dielectric constant and shallow depth of objects.

The ground radar has been successful in locating non-metallic pipes, cables, filled-in mineshafts, and underground hollows like adits, tunnels and caves. It can find metallic and non-metallic objects, as well. It is most suited to investigating

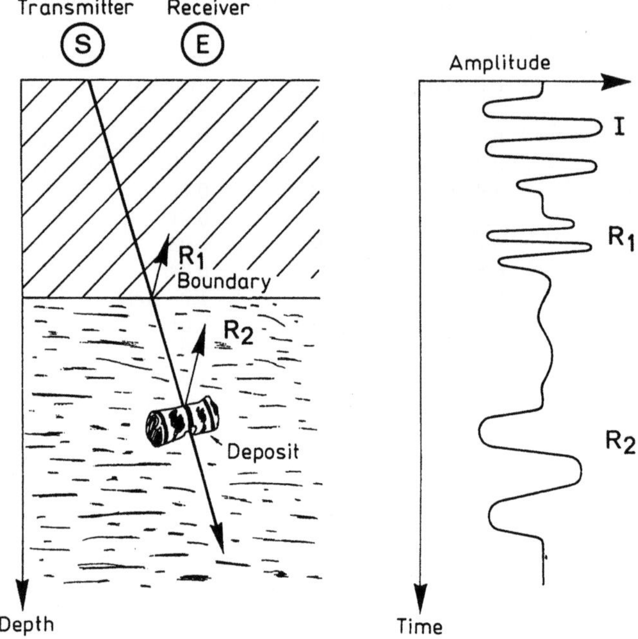

Fig. 2.17. Principle of ground radar measurements

abandoned industrial and military sites, where contaminations occur in shallow ground.

To achieve more depth penetration, according to the nomogram of Fig. 2.12, the transmitted frequency should be lowered, although lower frequencies (<100 MHz) decrease the resolution. The frequency must therefore be chosen by considering both the possible depth penetration and the desired resolution. In addition, the electric properties of the survey area and the targets of the investigation must be weighed.

Ground radar surveys result mostly in a great number of reflections, which pertain to small alterations in the textures and structures of bedded soil. It is presumptuous to interpret such patterns as indications of contaminated waste: the evaluation and the assessment of radar data must consequently be done with special care and precaution.

Electromagnetic Sounding

It has already been stressed that a reduction of frequency increases the depth of penetration (Fig. 2.12). This effect is used to carry out multiFrequency Electro-Magnetic (FEM) Soundings. As many frequencies as possible between 1 and 60 Hz are sent into the earth. A receiver measures the horizontal (H_r) and vertical

(H_z) magnetic components of the stationary electromagnetic field for all transmitted frequencies.

The EM field is induced by a big cable loop, acting as a coil, which is horizontally laid on the surface. During one set of observations, transmitter and receiver remain fixed. Finally, the ratio of the two magnetic components H_r/H_z and their phase differences $\Delta\phi$ are registered.

The digital evaluation of the field data is complicated and can only be achieved by using computers or PCs with large memory capacities. The interpretation leads to results, which are comparable to the DC depth soundings. However, FEM sounding is more expensive and should be employed when DC sounding fails. This may happen because of too high or too low resistivities of overburden.

Divining Rod

The divining rod has been used since the Middle Ages to localize strong minima of the electromagnetic field, which are narrow and extended in the direction of strike. Nowadays, the rods are applied to locate metal pipes and cables in shallow ground. Most in use is a primitive instrument that consists of two welding rods whose ends are bent by 90°. When the rods are held forward and horizontally by hand, they will, according to Lenz's law, turn in the direction of the electrodynamic field, while the diviner crosses an EM anomaly.

This archaic electromagnetic method requires very dry palms and a steady progress of the diviner. The error rate is very high. Metallic pipes, for instance, must be embedded in sand or other material of high resistivity to move the divining rod. It is safer to use a commercial electromagnetic metal or cable detector. Statements regarding depth and extension of aquifers or of geological structures are mere speculations. Costly follow-up work, like drilling, should therefore not be based on divining.

2.3 Seismic Methods

2.3.1 General

Seismic investigations are based on different elastic properties of the rocks of the upper crust. A seismic wave can be created on the surface by artificial seismic sources like hammer, weight-drop, vibrator or explosive charges. The wave runs through the earth with a velocity that depends on the traversed rocks or materials. At interfaces where the seismic velocity or density changes, seismic waves are diffracted, refracted or reflected.

Portions of the primary wave return to the surface after travelling different distances through the ground. There, the remaining seismic signals are registered by a number of seismic receivers, called geophones. Mostly, they are arrayed on a single line, but other arrays may also be used. By evaluating the travel time between the break and the recording of a seismic signal, the seismic velocities of strata, their location and the depth of their seismic reflectors may be inferred.

Consequently, seismic work will provide special knowledge about the thickness and extension of layered strata and structures of the earth, which is essential to solve geological or hydrological problems. In addition, the shape and extension of waste deposits can be comprehended.

The propagation of seismic waves follows the geometry of optical laws. The refraction of seismic waves at the boundary between two beds with lower velocity at the top and higher velocity at the bottom layer is described in Sect. 2.3.2 "Seismic Refraction" (Fig. 2.18). In contrast, the "seismic reflection", delineated in Sect. 2.3.3 and in Fig. 2.19, is based on the reflection of seismic waves at many stratigraphical or structural boundaries.

At every seismic break, artificial or natural, the ground is submitted to compressing and shearing processes. Accordingly, different types of elastic waves are emitted, which travel at different speeds. A seismic wavefront expands from the source in all directions through the rocks in the ground. Two types of waves are discerned: the compressional or longitudinal wave and the shear or transversal wave. Compressional waves travel faster than shear waves and reach their target first. Therefore they are named primary or P-wave, the slower shear wave is called secondary or S-wave.

The particle motion in S-waves is perpendicular to the direction of propagation. Due to their smaller velocity, they provide a better resolution of structures. They cannot, however, penetrate liquids and are rapidly weakened when travelling in loose sediments. The registration of this wave type is more difficult than for P waves.

Besides these waves, which propagate through the ground, surface or interface waves run along the surface of the earth and may travel by several modes. The most common of these are Raleigh waves. Other surface waves include the fast love waves, hydrodynamic waves and Stoneley waves. If they travel in a borehole, they are called tube waves. Surface waves often disturb the seismic signals of near surface layers.

Estimation of seismic velocities has to observe the following rules:

1. Inside most layers, the velocity increases continuously by growing depth. This effect is especially strong near the surface, since the consolidation of strata progresses there more rapidly.
2. Normally, the seismic velocity is faster in the underlying than in the overlying layer.
3. Seismic velocities depend mostly on the geological age of the strata. The rule is: the older the bed the higher its velocity. This does not apply to near surface beds. Here reduction of pressure and weathering have led to too small velocities.

2.3.2 Seismic Refraction

Seismic rays are refracted upon passing into a bed with higher velocity. The waves travel along the interface of the two beds and continuously emit seismic energy to

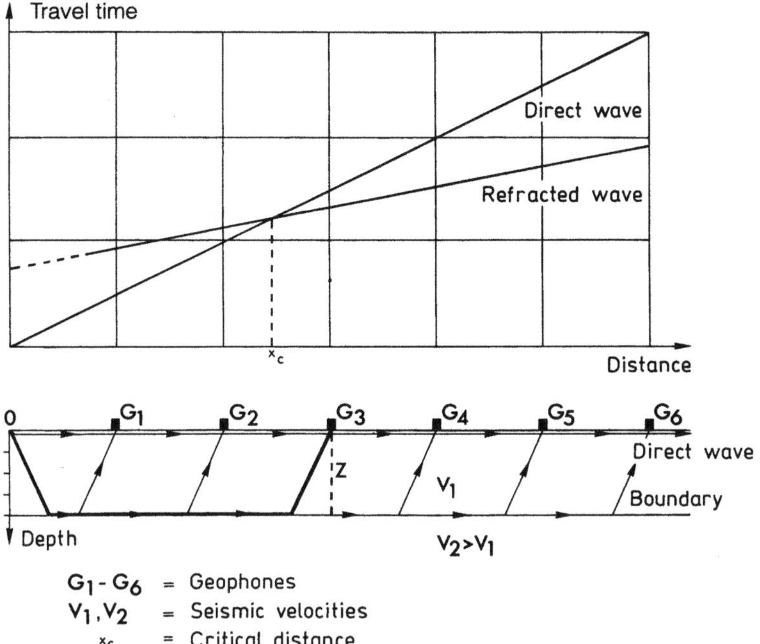

Fig. 2.18. Principle of seismic refraction

the surface. The precondition is that seismic rays hit the interface at a critical
angle.

A schematic picture of seismic refraction is presented in Fig. 2.18. The lower
diagram describes the travel path of the waves, which run directly and by refraction. The upper diagram displays the travel time curve, which contains the registered data. The x-axis is divided by the distances between seismic source and the
geophones, the vertical y-axis is marked by the travel time. Fig. 2.19 shows
graphically the procedure of refraction field work and the steps of digital evaluation.

At the earth's surface, geophones register the refracted waves and, in addition,
the direct wave, which travels along the upper layer. Since the refracted wave
moves with the greater velocity of the lower layer, it will overtake the direct wave
at a critical distance X_c from the seismic source point. This distance is a measure
of the depth of the boundary between the two beds. The seismic velocities may be
inferred by complex mathematical evaluation.

The total length of a seismic string and the geophone interval determine the
penetration. As a rule, the string should be at least five times longer than the desired penetration. In case only shallow penetration is requested, the geophone
interval can be reduced to a few meters only.

Seismic refraction is especially geared for the investigation of boundaries in
shallow depth. The reflection method is not so well suited for this task. Its targets
are found in depths of > 50 m.

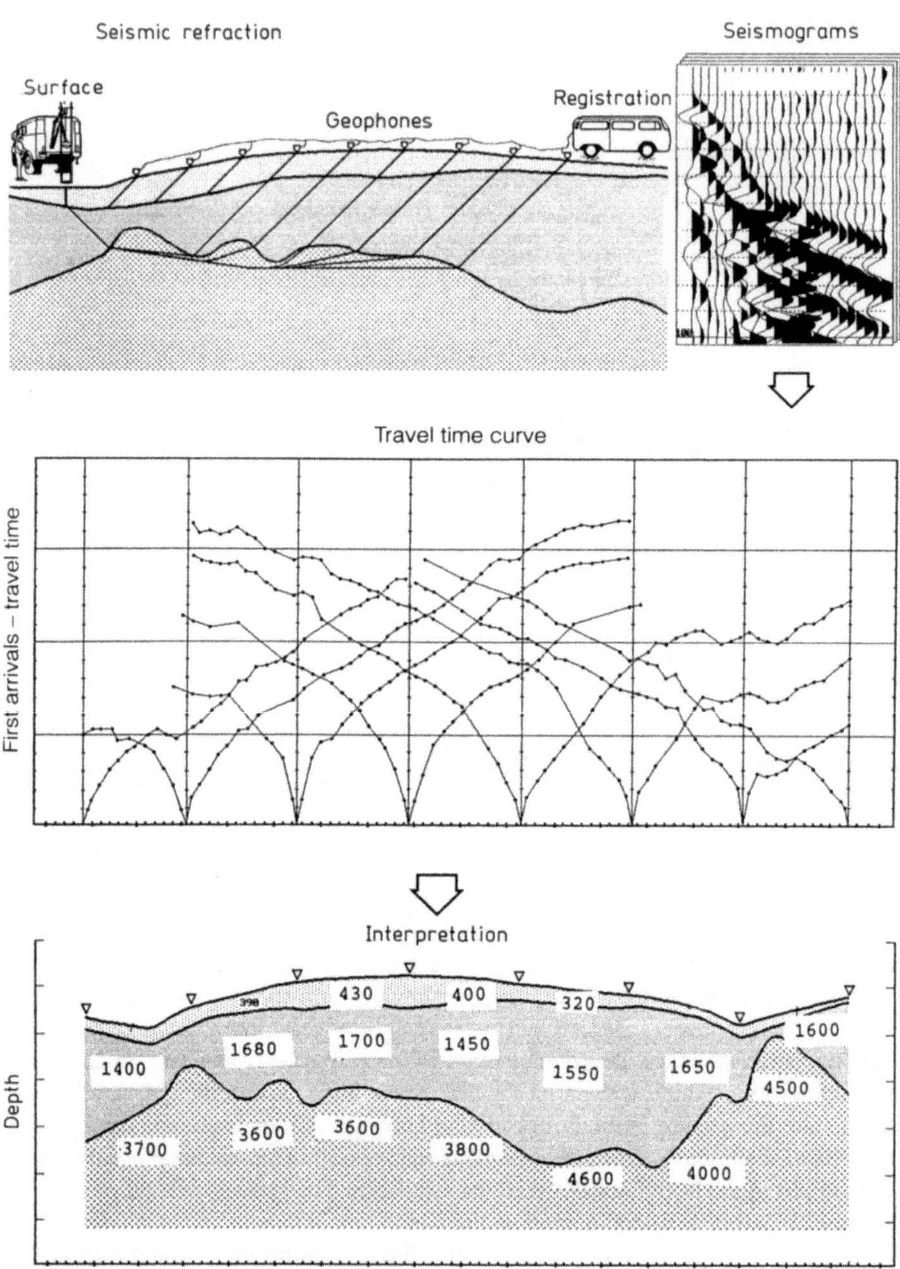

Fig. 2.19. Pattern of seismic refraction

Figure 2.19 portrays at the top the refraction array in the field, the survey cars and the raw seismic record. The middle section depicts the evaluation through the construction of the travel time curves. The final result is presented at the bottom, where seismic velocities and the depth of seismic reflectors are combined.

The seismic refraction method is well suited to investigate the geological beds at the bottom and in the vicinity of hazardous waste dumps. Good results have been obtained by the determination of the thickness of unconsolidated rocks or of the weathered zone. Detailed isoline maps of the surface of impenetrable beds (aquiclude) can be constructed by refraction data. Even ground water guiding channels can be mapped, because they are filled with sand and gravel of low velocity. They show clearly the paths any leachates or contaminated plumes will follow.

Furthermore, seismic refraction provides knowledge about the extension, thickness and depth of hazardous waste dumps, provided seismic velocities distinguish well between dump material and the country rock. This is the case when, at the bottom of a waste-deposit, hard bedrock is present. However, a distinction between waste dump and unconsolidated or weathered rock is mostly impossible.

2.3.3 Air-Acoustic Seismics

Seismic exploration of shallow targets is hampered or made impossible by strongly varying conditions of propagation, which prevail directly under the surface of the earth. Therefore, it is advisable to use several modified seismic signals. But customary seismic sources mostly produce only one type of signal.

To employ seismics also in shallow ground, the air-acoustic method has been developed. Its signals can be varied over a wide range and thus be adapted to the seismic properties of soil and shallow strata.

Electronic signals of different wave forms are created by frequencies between 50 and 100 Hz. The seismic source is a big loudspeaker, which is placed upside down on the surface to project the sound waves into the ground. The thus created seismic waves are received by a conventional geophone.

The loudspeaker source is easy to move; the air-acoustic survey is therefore performed in a mapping array with a constant distance between loudspeaker and geophone (Fig. 2.20). As with other seismic methods, alterations of travel time indicate the presence of discontinuities. They may be caused by buried material or by reflecting layers.

2.3.4 Seismic Reflection

Seismic waves are produced by a source quite similar to the seismic refraction, which is mostly located on the surface of the earth. At discontinuities in the ground, i.e. horizons at which the seismic or acoustic impedance is altered, however, they are not refracted but reflected, and the reflected waves are received by a string of geophones on the surface.

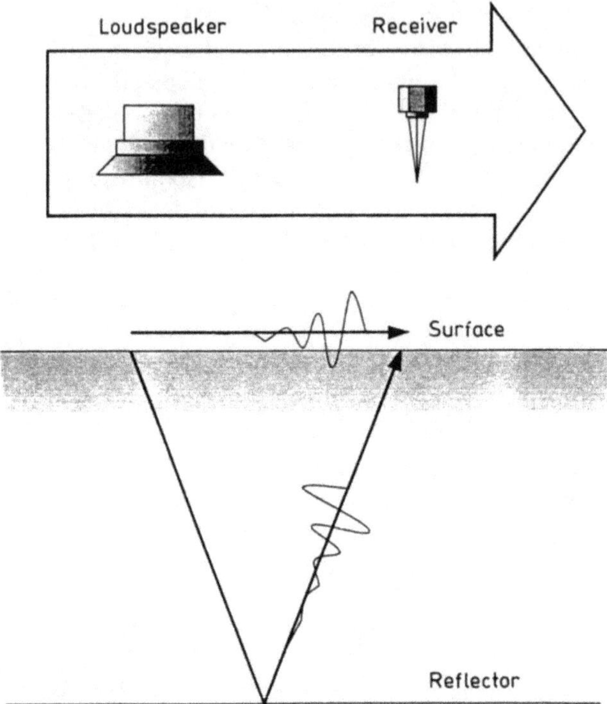

Fig. 2.20. Principle of air-acoustic seismics

The seismic impedance is the seismic velocity multiplied by the density of the strata. This implies that seismic reflectors may occur at the interfaces of beds with different lithology. It is important to know that these horizons must not agree with changes of the specific electric resistivity. Therefore, discussions of different depth estimates by seismic and geoelectric methods are futile.

The reflected waves, which have been registered at the surface, mostly along survey lines, are converted by geophones into electric signals, which are evaluated, formerly by graphic methods but nowadays by computing the travel times, into seismograms. They already roughly display the structural pattern of the sediments, which contain seismic reflectors, but to calculate the depth of strata, the seismic velocities have to be known (Figs. 2.21 and 2.22).

Figure 2.22 displays at the top the field array for seismic reflection with two cars. One acts as the seismic source by means of a huge vibrator, which is installed inside. The other car is a mobile station for registration, where the seismic signals of the numerous geophones are filtered, stacked and stored on magnetic tapes or disks.

In the middle, the outcome of filtering, stacking and other means of evaluation is shown as a seismogram. The final result appears at the bottom, where reflectors are coordinated with the boundaries of geological beds, according to their travel times. A plausible picture of the geological structure of the section is drawn.

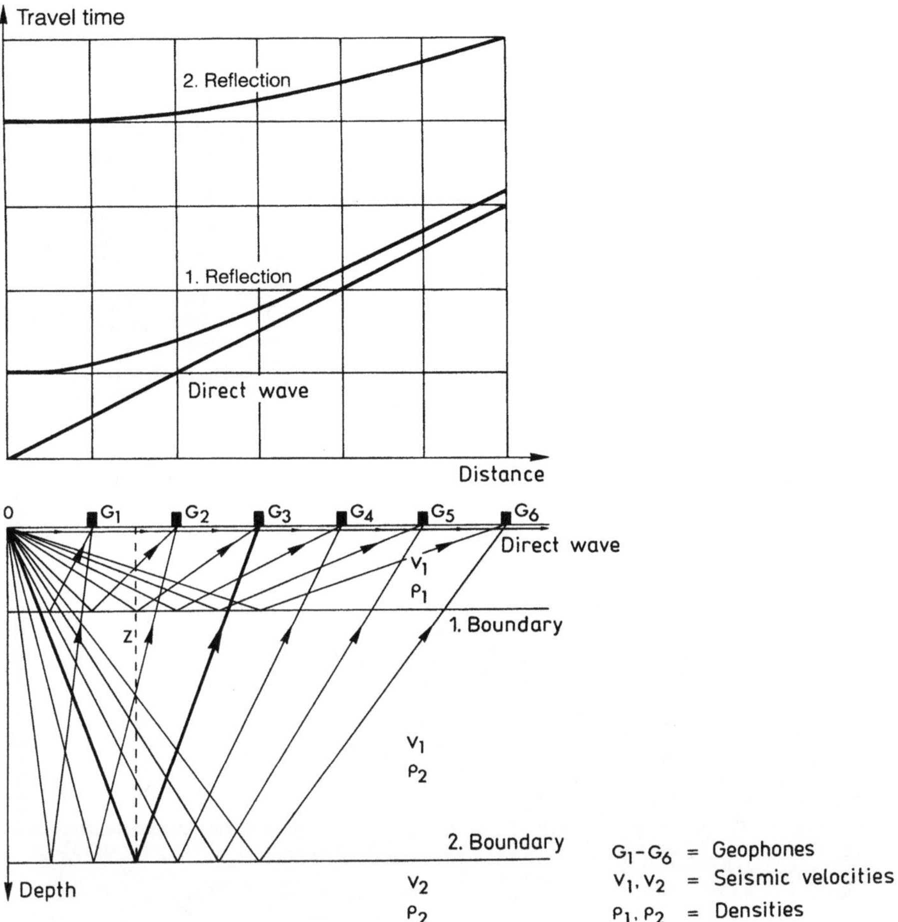

Fig. 2.21. Principle of seismic reflection

An advantage of seismic reflection is the short strings of geophones. Since the geophones can be positioned near the seismic source, a greater depth of investigation can be reached than in refraction surveys, using the same length of string.

Hitherto, the realm of reflection work was oil and gas prospecting, which dealt with geological structures in depths from several hundreds to several thousands of meters. The seismic field and evaluation technology was developed to perfection to explore this deep.

By contrast, the application of seismic reflection to near surface objects, like hazardous waste dumps, is still under development. The special difficulty that must be overcome is that reflected seismic signals return to the surface or to the geophones so fast that Raleigh and related waves are not yet attenuated and reflected signals will be suppressed.

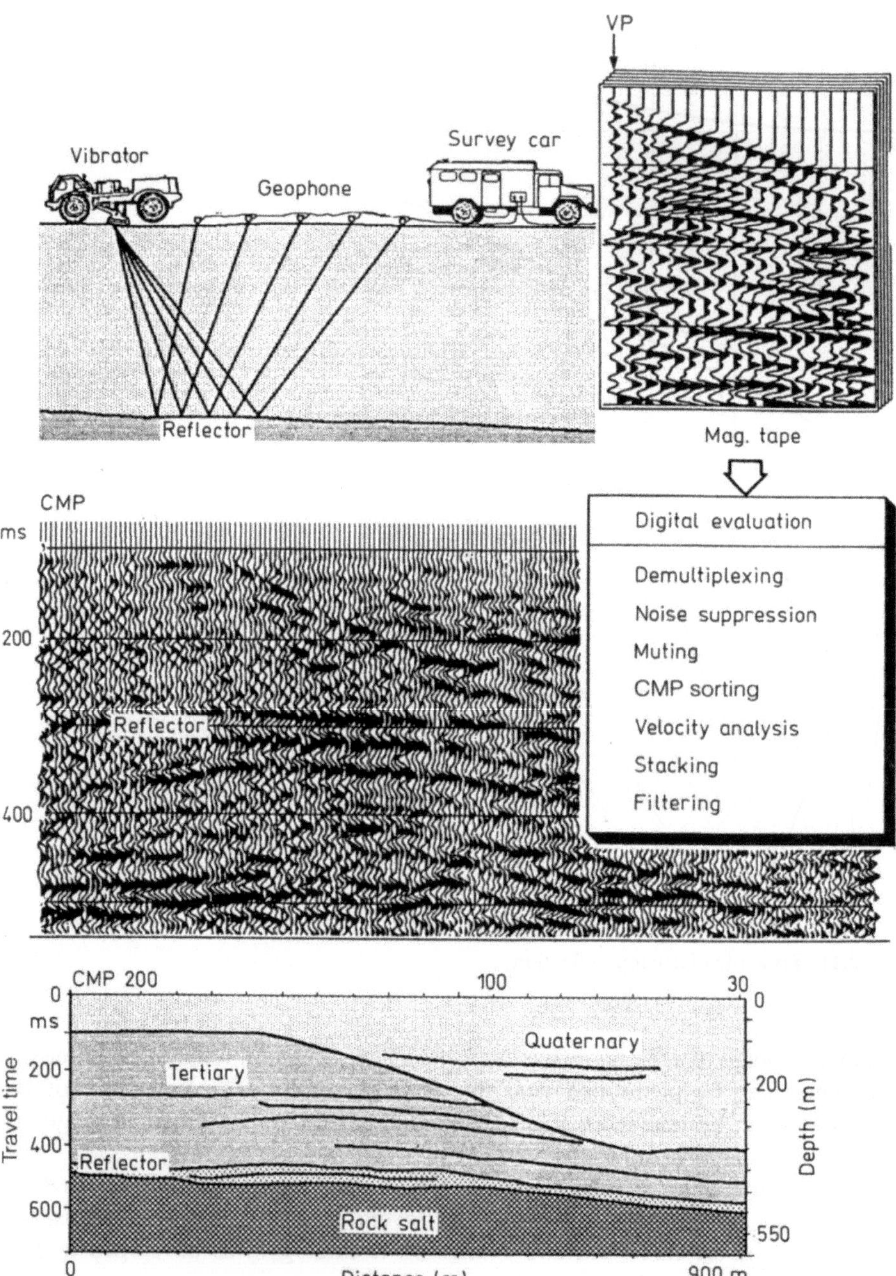

Fig. 2.22. Pattern of seismic reflection

To make reflections from depths $<50\,\text{m}$ visible, seismic receivers with extremely high sampling rates and high-frequency sources have to be used. This high-tech equipment is now available and it is possible to obtain seismic reflections also in polluted areas near the surface. One should bear in mind, however, the high costs of seismic reflection when planning such a survey. In many cases, the cheaper refraction or geoelectrics will fulfill the same purpose.

Often, information about deeper layers is desired, even for the solution of environmental problems. For example, the structures that control the flow of contaminated plumes should be followed to greater depth to assess the danger of widespread underground contamination. In such cases, reflection surveys are well suited to trace stratigraphically and/or tectonically marked seepage paths.

The evaluation of digital registered reflections has to deal with huge amounts of data. The following steps are customary:

1. Editing (control of field values)
2. Demultiplexing (lines $\rightarrow$ columns)
3. Correction of amplitudes ($\rightarrow$ common level)
4. Static correction (topography, overlayers)
5. CDP sorting (reference to common midpoint)
6. Stacking (summation of singular seismograms)
7. Deconvolution (elimination of multiple reflections)
8. Bandpass filtering (elimination of noise)

After the evaluation follows the digital interpretation, which implies the processing of great volumina of evaluated data. It can be done only by very fast computing systems with large memory capacities. The purpose is to extract only such waves from a great number of uncoordinated oscillations as pertain to reflections at the boundaries of layers. Out of many processing procedures, only the migration can be mentioned. It is the computation of a theoretical wave front, which exists at the moment of a seismic break.

2.4 Gravity

Gravity surveys are based on the variations in the gravity field of the earth, which are caused by inhomogenities in the density of structures in the ground. To measure these density anomalies, raw field data have to be intensively corrected by known reference values, which depend on the time and location of the survey. These are: influence of tides (isostatic correstion), elevation above a reference level (free air correction), topographical relief (terrain correction) and density/ thickness of rocks that are situated between survey and reference level (Bouguer correction).

These corrections may be larger than the mostly weak gravity anomaly of a hazardous waste dump. Faulty corrections, which can originate from a wrong estimation of the density of dumped material, may lead to a false interpretation.

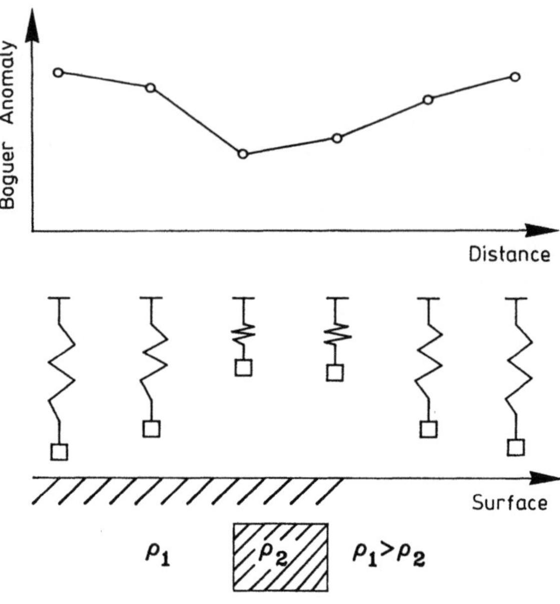

Fig. 2.23. Principle of gravity measurements

The pattern of gravity surveys is displayed in Fig. 2.23. Gravity measurements are made by gravimeters, which are basically highly sensitive spring scales. The alterations in the length of the spring are directly related to changes of gravity. The production of gravimeters demands the greatest precision and vast mechanical skills. Gravity is measured by a unit of gravity acceleration equal to 0.1 mGAL (milligal) or 10^{-6} m/s^2.

The application of gravity surveys to environmental problems is limited because of high cost and small gravity anomalies. For example, the surveys need to be prepared by precise levelling to eliminate the strong influence of minor topographic features.

Some gravity investigations of hazardous waste dumps have nevertheless been successful. In cases where a strong density contrast existed to the country rock, they were able to find the border of the dumps and to outline the walls of sealed basins. Gravity measurements are well suited to locate underground cavities. For this, very dense (and expensive) grids of gravity stations have to be surveyed.

2.5 Geothermometry

Geothermal surveys from the surface of the earth consist of shallow measurements of temperature to map geothermal anomalies. Two methods are customary:

- Infrared (IR) surveys of the surface.
- Temperature measurements in flat boreholes or in probing holes from decimeter to few meters depth.

The IR surveys gauge the temperature of the surface by contactless and mostly airborne infrared detectors in thermal scanning devices. This method is expensive and susceptible to all changes of weather, especially to the duration of sunshine. Its applicability to environmental problems is therefore restricted.

The measurements in percussion probe holes or flat drill holes are made by resistance thermometers, which are housed in the pointed end of special probes. The temperature is transferred to the sonde by heat conduction. Such observations should be performed only after the distortion of temperature caused by probing or drilling has ceased. Normally, one should wait ten times the length of time that was needed to sink the hole.

To eliminate the falsifying influence of sun rays, temperature surveys should be performed only at night, preferably between the hours of 4 and 6 a.m. Strong rain or periods of strong evaporation should not precede temperature observations, since they may alter the values considerably.

The application of geothermic work to the elucidation of environmental problems like finding hazardous waste is limited. The reason is the too-small density of heat flow on the surface. Even where strong heat was created by microorganic decay of cadavers inside dumps, only temperature anomalies, which were weaker than those influenced by the weather, were found on top of disposal sites.

A proper temperature survey has therefore to be very precise. It is necessary to monitor the diurnal variations of temperature at a base station continuously, and to correct the raw values accordingly. Needless to say, the temperature base should be installed at the same depth as the field measurements. The survey grids should be small-meshed; the rectangular distances of stations should be < 5 m. Since leachates or seepages coming from hazardous waste dumps are normaly slightly heated, they may be traced by temperature determinations in flat boreholes > 3 m deep.

2.6 Well Logging

2.6.1 General

By geophysical well logging the physical properties of rocks, ground water or disposed material are determined. The findings are valid only for the direct vicinity of the hole. Every borehole should be logged, since this provides the best key to the proper interpretation of geophysical field data and helps in the recognition of contaminations from the surface.

In this chapter, an overview of the customary well-logging methods is given. It is not necessary to run all the described logs in one borehole. The reader should choose those combinations that will best suit his special environmental problems.

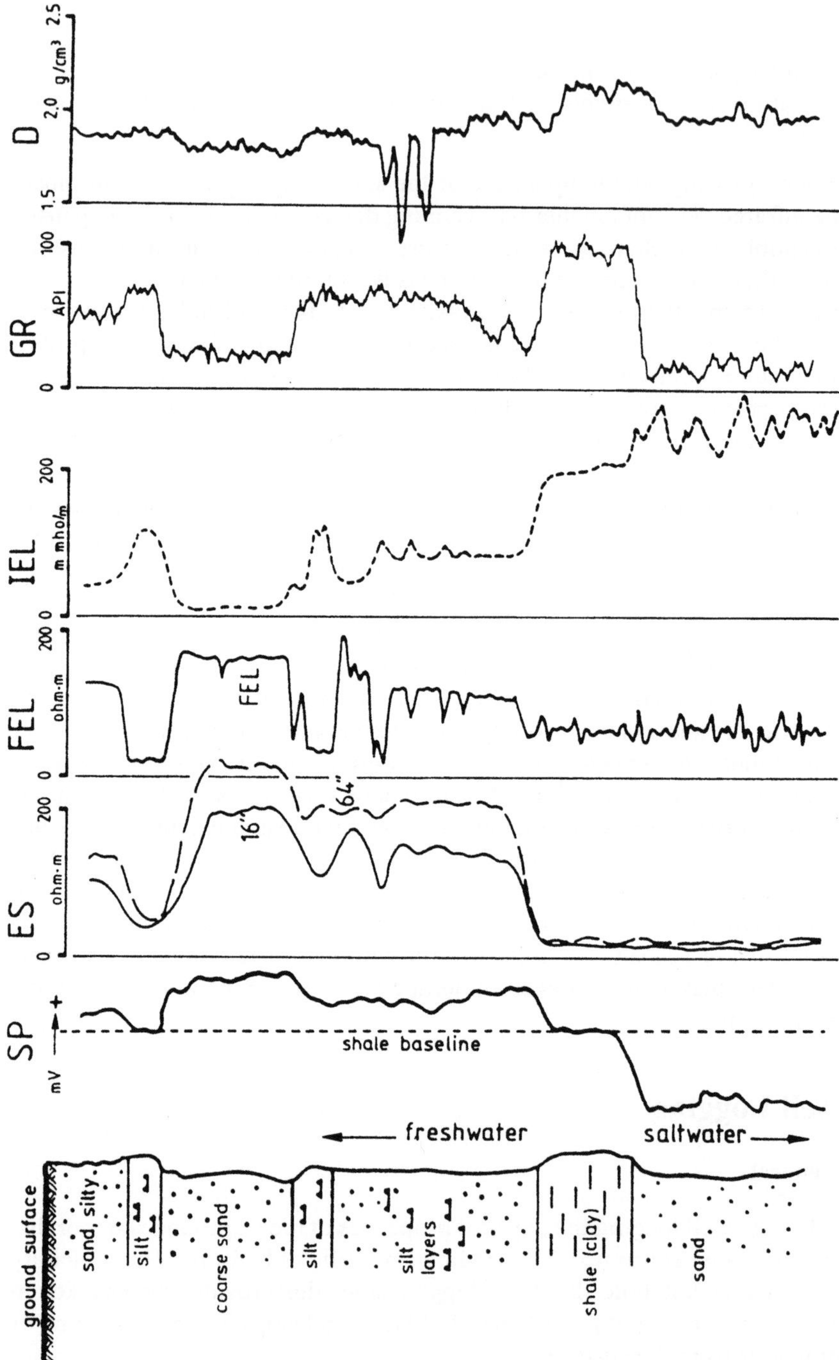

Fig. 2.24. Comparison of different logs with the lithology of cores.
SP = self-potential survey; ES = electrical survey measures resistivity in 16″ and 64″ point array; FEL = focused electrical log for thin layers; IEL = induction electric log measures electric conductivity; GR = gamma ray measures natural radiation; D = density log by artificial gamma source and detector

Log data are recorded continuously while the probe moves in the hole with constant cable speed. This speed is limited by the necessary resolution and the time needed for one measurement. Fig. 2.24 presents examples of complex well logging.

2.6.2 Logging Methods

Gamma Ray (GR)

The gamma-ray probe measures the natural gamma radiation, which originates from the potassium isotope ^{40}K and from the uranium and thorium decay series. This log enables the discrimination between layers of clay and sand; even the clay content of clayish sediments can be estimated. Barrages by impenetrable minerals may also be controlled by gamma-ray logs.

The probes contain scintillometers and can be operated in dry or cased boreholes. The logging speed should not exceed 5 m/min. Beds of only 0.3-m thickness may still be resolved.

Density Log (D)

This is also a radiation log, but here variations of the radiation of an artificial source are registered. This source (^{137}Cs, Cäsium 137) is placed at the low end of the probe. The gamma detector is housed above, shielded by a lead column against direct radiation from the source (Fig. 2.25).

The gamma radiation that is emanated into the surrounding rock is absorbed by the latter, according to their density, by the Compton effect. The part of radiation that still reaches the detector is recorded and is a measure of the rock density.

The source-to-detector distance is conventionally 40 cm. Then, a horizontal expansion of roughly 15–20 cm is reached. The density log is applied to differentiate among several dumped materials, to determine their boundaries, and to locate fractured seepage paths in consolidated rocks.

Neutron Log (N)

An artificial neutron source (mostly americium-beryllium) radiates in the borehole fast neutrons. They collide with atoms of the drilled rock and thereby lose energy. When they reach a certain level of energy, some are caught by other nuclei. These are excited to higher energy radiation.

Detectors that are arranged at the probe at some distance from the neutron source measure the captured gamma radiation and/or the thermic neutrons. Needless to say, the detectors must be shielded against natural radiation.

Neutron logs depend on the hydrogen content. Therefore, water and hydrocarbons are encountered as pore fluids. After borehole effects of diameter, mud, etc., are corrected, the rock porosity and permeability can be estimated.

Electric Log (EL, ES)

This log determines the apparent resistivity of rocks in multiple point arrays. Most probes allow also the simultaneous registration of the self-potential between

Table 2.4. Logging methods, measured parameters and objects of investigation

Symbol	Parameter	Result	Object
GR	count of natural gamma radiation	natural radioactivity of rocks	petrography clay content
D	counts of compton scattered rays	density of rocks	fracturing, porosity
N	counts of secondary neutron-neutron rays	lithology	stratigraphy porosity
EL, ES	apparent resisitivity	true resistivity	hydraulics, lithology
ML, MLL	apparent resisitivity at borehole wall	true resistivity small scale	lithology, hydraulics
IEL	app. conductivity, focused induction	true conductivity	lithology
FEL, LL	focused electric log	true resistivity of rock	lithology
SP	self-potential (probe-to-surface)	sources of electric potentials	oxidizing bodies
SAL	resistivity of borehole fluid	salinity	total salt content of fluid
TEMP	temperature of borehole fluid	geothermal field	thermal gradient
SONIC SV	travel time of seismic waves	seismic velocity	seismic velocity
CAL	borehole diameter	shape of borehole walls	correction of other logs
FLOW	revolutions of a spinner	velocity of fluid flow	zones of in- and outflow of water
DV	compass and dipmeter	inclination + azimuth of borehole	spatial drill path
OPT	video signals, photography	state of borehole walls	direct view of lithology

a surface current electrode B and a potential electrode M in the hole. The measured apparent resistivities, also called mixed resistivities, are a combination of the electric parameters of the borehole mud, the mud cake, and the flushed and invaded zones of the rock. Figure 2.26 displays these conditions. Therefore, the apparent resistivity must be converted into the true rock resistivity via correction programs or departure curves.

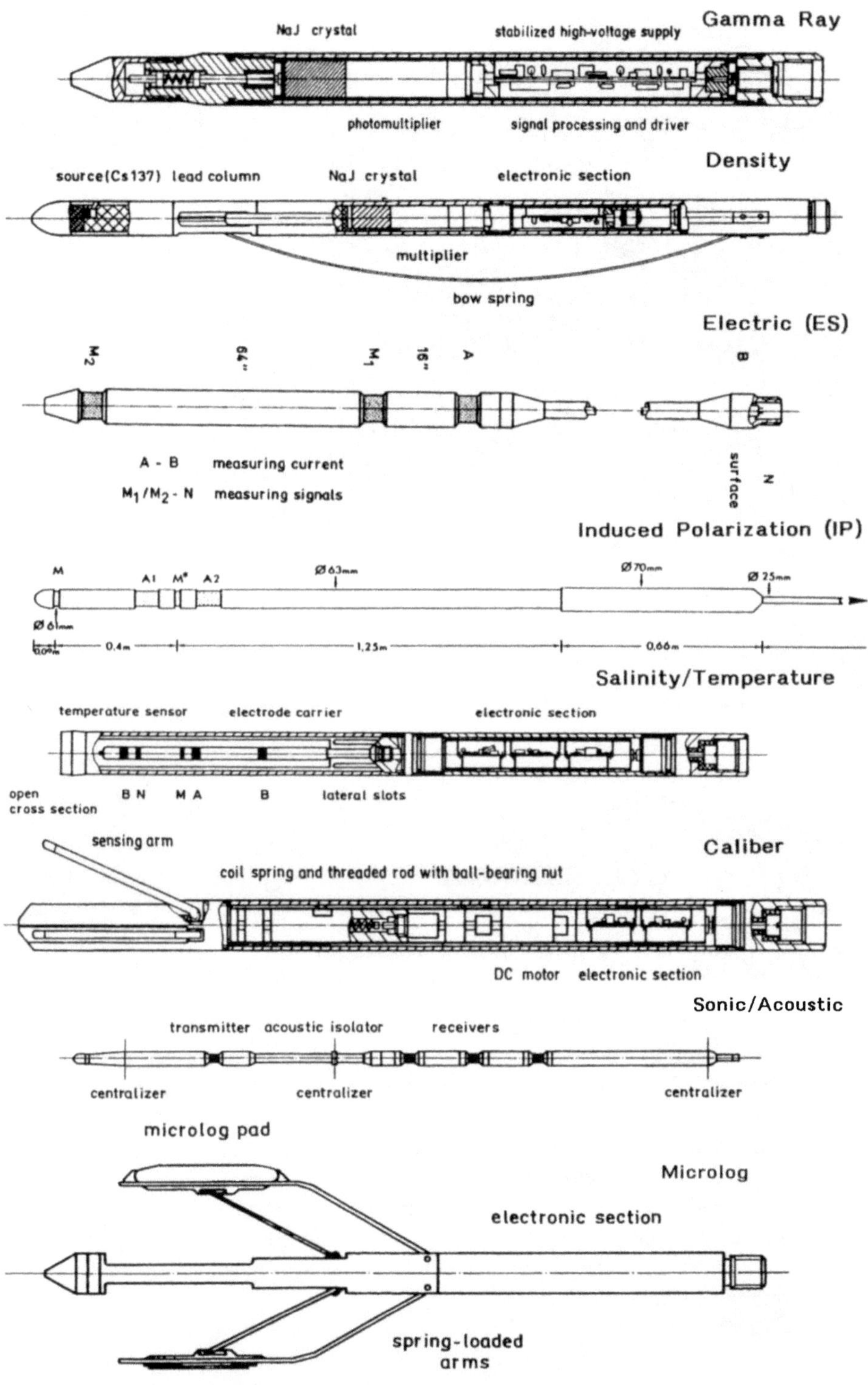

Fig. 2.25. Probes for geophysical well logging

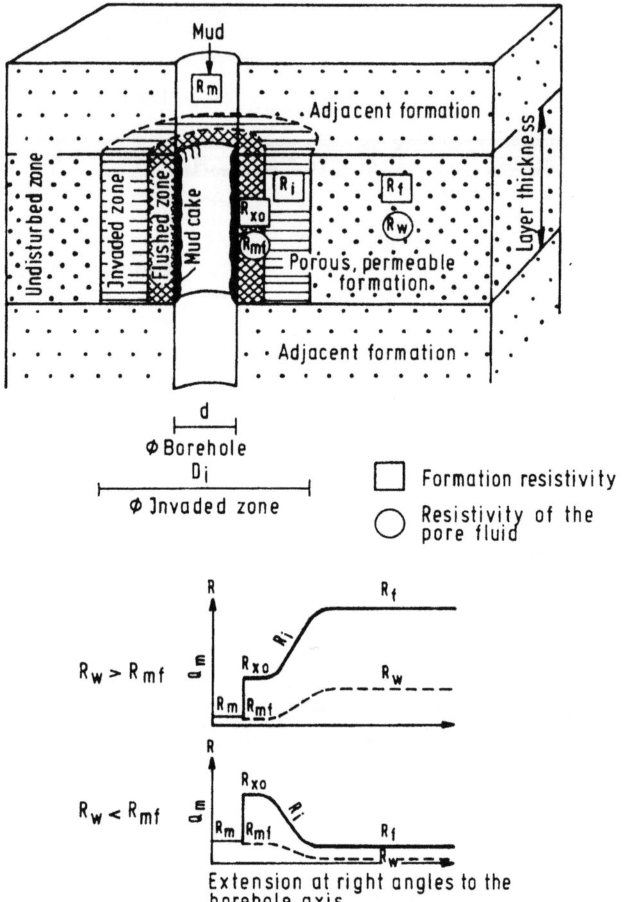

Fig. 2.26. Distribution of resistivities around a borehole

The most common ES electrode array is the combination of the 16″ normal and the 64″ normal (Fig. 2.25). The shorter has a small penetration and is strongly influenced by mud and its infiltration into the surrounding rock. However, it resolves even small layers down to 0.5-m thickness. The longer spacing reproduces the rock resistivity better, but can resolve only layers >2 m.

ES logs should always be run in environmental drilling programs, since the measured resistivities provide knowledge about the sequence of deposits, the depth and salinity of leachates and the clay contents of basal or top sealings, if perforated by drillholes.

Monoelectrode Log (PR)

Other than the multipoint ES log, a single-pointer electrode array may be used. It is the simple determination of the resistivity between an electrode on the surface and another at the probe. It has a fair resolution of thin beds but it is strongly influenced by the mud and the invaded zone, and therefore unsuitable to detect the rock resistivity. Whereas ES logs are non-focused and of simple technology, better resolution and deeper penetration is reached by *focused methods*; these, however, require more complicated electronic guidance.

Microlog, Microlaterolog (ML, MLL)

The resistivity distribution in the immediate vicinity of a drillhole is measured by an array with electrode spacings of $1-2''$. The electrodes are pressed hard against the borehole wall while moving; the mud is pushed away and the mud corrections can be omitted.

The small electrodes may be arranged in the unfocused microlog or in the focused microlaterolog or similar arrays. The purpose of this fine-logging is to resolve very thin layers, to locate single fractures, joints and fissures, and to determine the resistivity of the mud-flushed volume.

Induction Log (IES, IEL)

This log, which is also known as the focused induction log, determines the reciprocal of resistivity: the conductivity $1/\Omega m$. Its unit is the mho/m or S/m (Siemens/ meter). Electromagnetic waves of the frequency 20 kHz are transmitted by a coil on a probe. This causes eddy currents in materials or rocks of different conductivity. They are received by another coil, which is located ~ 1 m away. Rock conductivity is calculated from amplitudes and phases of the received secondary field.

The advantage of IES is that it can work out true rock conductivities of very low resistivities. It is suitable for the exploration of rocks that are infiltrated by saline fluids or leachates.

Focused Electric Log, Laterolog (FEL, LL)

The electric current of one borehole electrode is focused to resolve beds, which may be as thin as 0,2 m. The measuring electrode length of $4''$ is carried by a probe of the total length of 2 m with the small diameter of $\sim 14''$. In spite of this minute array, its lateral penetration reaches as far as the $64''$ normal. This is achieved by placing additional electrodes above and beneath the current electrode.

However, these advantages of FEL are accompanied by drawbacks. Higher resistivities cannot be recorded and the resolution is reduced by larger hole diameters and by mud invasion.

Self-Potential (SP)

Such electric potentials are composed of electrochemical and kinetic potentials. They may originate from dumped metals, or sulfidic ores, which undergo oxidation or reduction processes, or from fast-moving gases or liquids. SP can also be caused by the metal casing.

In principle, it is valid that positive SP hints at increasing salinity of the pore fluid and that negative SP may indicate fresh water, provided the clay content remains stable within the logged depth.

In combination with ES, GR and IP logs, estimations of the clay percentage and the permeability of strata can be attempted. A precondition for a quantitative interpretation is a clearly marked sand-clay interbedding and different salinities of mud filtrate and pore water.

Salinometer (SAL)

The salinometer probe measures the specific resistivity (salinity) of the borehole fluid. The electrodes are closely spaced and housed in an insulated metallic tube, through which the borehole fluid passes. This arrangement is made to prevent the influence of the resistivity of the drilled rock.

Knowledge of the salt content of the borehole fluid is necessary to calculate the true resistivities of rock from various electric logs. In gauge wells or boreholes sunk through hard rock, the salinity can indicate where water inflows or outlets occur. The salinometer probe is often combined with a temperature tool and is used on the down-hole run.

Temperature (TEMP)

The temperature of the borehole fluid is continuously monitored by an electric resistance thermometer in relation to depth. The accuracy of the measurement reaches 0.01 C (degrees centigrade). To obtain undisturbed temperatures, one has to wait until the disturbances caused by drilling have subsided.

Because of the annual temperature variation, the natural depth increase can first be observed from 20 m downwards. The geothermal gradient normally averages 3 °C/100 m. Deviations from this can indicate ground-water movements. This is very pronounced at water inflows; there, the temperature log shows sharp kinks. Especially at hazardous waste dumps, temperature anomalies in boreholes may indicate exothermal chemical or biological reactions between the deposited materials which occur near the borehole.

Sonic Velocity (SV)

This is also known as acoustic or sonic log and continuously records the travel time of longitudinal waves between two points on the rocks of the borehole wall. The sound transmitter is housed in the probe, as well as one or more detectors. The measured travel time is related to the lithology and the porosity.

Sonic logging is difficult and expensive; the raw data have to be corrected for many effects, like cycle skips, borehole parameter, etc. Sonic probes contain a lot of electronic gear and are prone to disturbing influences.

The knowledge of sonic travel time or sonic velocity is important for the recognition of seismically active horizons and for the calibration of seismic refraction and reflection. Additionally, the porosity and the amount of fracturing may be determined.

Caliper (CAL)

Caliper tools measure the borehole diameter continuously. The important result is the deviation from the diameter of the drilling bit, which tells about cavings caused by loose sediments, frayed material or fractured hard rock. Narrowing of holes by swelling of clay or mud cake are also located.

Customary are caliper tools with three or four arms, which are pressed against the borehole wall while moving upward. Their spreading is recorded and is linearly connected to the depth. Caliper logs provide important data for the corrections of other logging methods and can locate cemented zones, casing and filters.

Flowmeter (FLOW)

This probe measures the vertical fluid flow in boreholes or wells. Its main purpose is to locate the depth at which ground water flows into a well. During a pumping test, the probe is lowered at a constant speed and continuously records the velocity of vertical flow.

Whenever it passes a water-producing depth, the revolutions will decrease. Thereby, the depth of the inflow is well marked. Flowmeter data, which are reduced by the cable speed, allow the computation of the share of the total production each producing layer has. This production can even be negative, if, for example, water flows out of the hole into a zone of fractured hard rock.

It is necessary to run an additional caliper log in hard rock to aid the interpretation of the flowmeter results. Preconditions for flowmeter logging are that the borehole diameter be not much larger than the tool, and that the velocity of water flow be sufficient. The vertical flow velocity has to be converted into pumping rates by multiplication with the cross section of the hole, found by caliper.

Deviation (DV)

The purpose is to ascertain the deviation of the borehole axis from the vertical and its azimuth towards north. The dipmeter probe is often a multishot instrument, which registers the dip and the orientation by taking photographs at every sequence of still measurements. More advanced are continuously surveying dipmeter probes, which record the spatial geometry of the borehole axis.

Deviation tools with a magnetic compass are restricted to open holes and do not work in steel casing. In this case, the expensive gyrocompass has to be employed. The deviation of a borehole should be measured as often as possible during the drilling operation to allow for early recognition of an unwanted course of the drill.

Other Logging Methods

There are a number of specific tools that are rarely used in environmental logging. Just to mention a few: optical or video logging uses videocameras to view the borehole walls. Needless to say, this works only in a dry hole or in clear water as borehole fluid.

The borehole televiewer works also in mud. It is not an optical tool but the borehole walls are scanned by pulsed, narrow sonar or ultrasound beams in a helix while it moves upwards. From travel time and amplitude, information is gained about fractures, fissures and strata with a very high resolution. A disadvantage is the very slow progress of the tool, which makes it applicable only to special zones, which are disturbed by tectonics.

2.6.3 Percussion Probing

This method is also known as geoprobing, drop penetration or driving. Shallow holes are made by driving small steel rods by hammer or hydraulic pressure into the ground without rotation. This is pertinent only for fine-to-middle-grained unconsolidated beds like clay and sand or other loose material. Hard rock has to be drilled in any case by rotating a bit. The depth of such probing is limited. In most cases,<10 m are reached.

The probing rods have a deep groove, into which samples or small cores of the probed loose material are pressed by the percussion. When retracting the rod, the samples come to the surface for inspection and investigation (core probing).

In many cases, the pressure at the point of the rod is registered as a measure for the hardness of the penetrated material. Furthermore, it is possible to do even geophysical logging in those short probing holes. This is done by pressing another, smaller logging stem into the hole, after the percussion rod has been pulled out. Resistivity and gamma ray logs have been run in percussion holes. There is no doubt that here we have a cheap and fast process at hand to countercheck geophysical near-surface data.

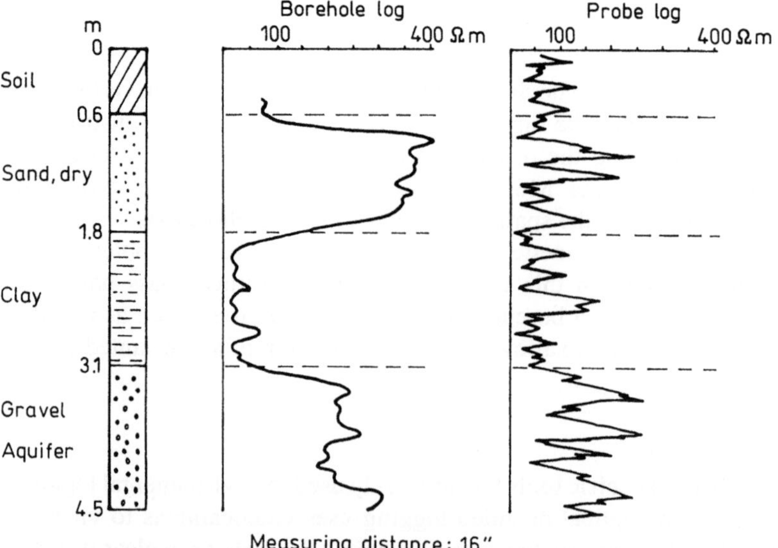

Fig. 2.27. ES (resistivity) logs run in a borehole and in a probe hole

The rare case in which resistivity was logged in a borehole and a percussion hole at the same spot is displayed in Fig. 2.27. In both holes, the resistivity was logged by the 16″ normal down to a depth of 4.5 m.

The differences in the resitivities of silt, clay and water-filled gravel range about 350 Ωm, but only in the borehole log are their boundaries so distinct that depths can be determined exactly. The percussion log has many more steep anomalous peaks, which permit no precise decision as to where the boundaries are situated.

2.7 Radioactivity

Chemical elements that possess more than 84 protons (atomic number $Z > 84$) are called radionuclides or radioatoms. They are instable, i.e. their atomic nuclei can disintegrate spontaneously under the emittance of radioactivity. During this process, the number of protons of the nuclei is changed and another chemical element is born. In addition, energy is produced, which is measured in electron-volts (eV).

$$1 \text{ eV} = 1.6 \cdot 10^{-19} \text{ joule (J)};$$

$$1 \text{ eV} = 4{,}45 \cdot 10^{-26} \text{ kilowatthours (kWh)}$$

The disintegration of atomic nuclei creates α-, β- and γ-radiation (Alpha, beta and gamma radiation). The α-radiation consists of helium nuclei; the β-radiation is composed of electrons and neutrinos that have no mass. It reduces the number of protons in the atomic nucleus by one. The γ-radiation, however, produces an extremely shortwaved and strong electromagnetic field. The wavelength of γ-rays ($\lambda \sim 10^{-11}$ m) is shorter than of visible light and of X-rays .

According to their very different nature, the three radiations possess varying penetration rates. The α-particles ionize other molecules on their direct way and lose their energy fast. They are not even able to go through a sheet of paper. Because of its lower atomic mass, the β-radiation has a higher velocity than the alpha particles; but for the same reason, the β-radiation path travels not straight, but zigzag. Only 3 mm of aluminium or 15 cm of clay shields this radioactivity. Most dangerous and most penetrating, though, is the γ-radiation. It penetrates matter quite easily and can be stopped only by a lead shield of >30 cm or by a 3-m thick layer of clay.

This means that radioactive waste cannot be detected by gamma-ray surveys if it is sufficiently covered by soil. In this case, the detection by the radionuclid radon (^{222}Rn) may lead to success. It is formed by the nuclear disintegration in the uranium-radium sequence, which starts from the most common uranium isotope 238 (Fig. 2.28). Radon has only a very short half-life period of 3.8 days. It moves slowly through the overburden and may finally escape into the atmosphere if it is not trapped by ground water. By determining the radon concentrations in the ground air, radioactive material, which lies much deeper, can be located.

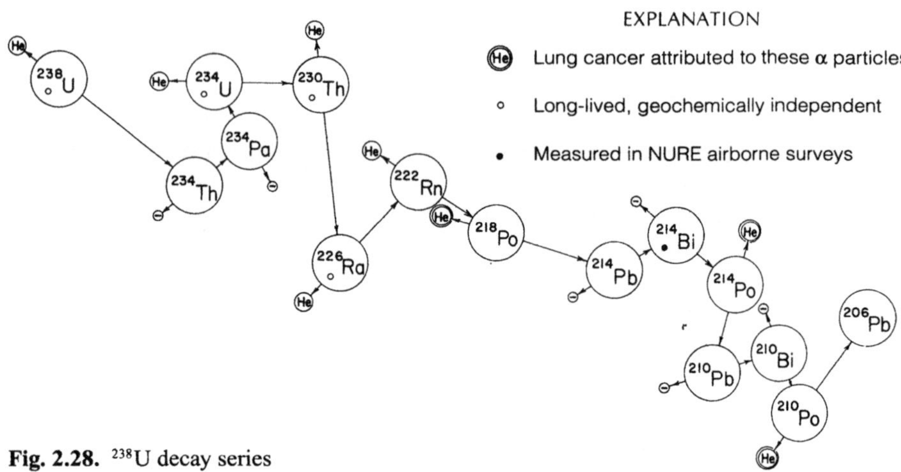

Fig. 2.28. ^{238}U decay series

The measuring unit of radioactivity is the becquerel (Bq), which is named after the French physicist who received the Nobel Prize in 1896 for the detection of radioactivity. One becquerel corresponds to one radioactive decay per second. Formerly the unit curie (Ci) was used (Table 2.5). 1 Ci equals $3.7 \cdot 10^{10}$ Bq. Therefore, even very weak radiation will be reported in large figures and the public may be frightened by an overestimated nuclear danger!

The number of atomic nuclei that disintegrate per second conveys nothing about the effectiveness of radioactive rays. The measure for this is the energy dose or the amount of absorbed radiation received by the radiated material.

Three kinds of radioactive doses are distinguished: the energy dose D, the ionic dose J, and the equivalent dose H. They are different for α-, β- and γ-radiation. The time dependence of the dose is expressed in units of dose performance (dose wattage).

The units of measurement presented in Table 2.5 are used mainly to describe and compare the damages inflicted by radioactive exposure on living creatures. Most important is the equivalent dose H, counted in mSv (millisievert), which gauges the dose of radiation a human body may encounter. The threshold values listed in Table 2.6 should not be exceeded.

The ionizing effect of radioactive radiation is used by the Geiger-Mueller counter to count the number of ionizations by the sudden collapse of a static voltage or potential field. Nowadays, a more precise and sophisticated method is favored. It is based on the property of crystals of sodium jodide to sparkle under radiation (radiophotolumincence). The brightness that is created is picked up by the photodiodes of scintillation counters.

It is worthwhile converting the scintillation values into Bq. The scaling into counts per second (cps) is less favorable, since this unit depends on the size of the crystal. Surveys by different instruments are then difficult to compare, unless they are calibrated in a test pit.

Table 2.5. Units of Radiometric Doses

Dose class	Mark	Definition	Unit		Conversion
			Old	New	
Energy	D	$D = \dfrac{E}{M}$	rad	Gy (Gray)	$1\ \text{Gy} = 1\,\dfrac{J}{kg} = 100\ \text{rad}$
Ionic	J	$J = \dfrac{Q}{m}$	R	$\dfrac{C}{kg}$	$1\,\dfrac{C}{kg} = 3876\ \text{R}$
Equivalent	H	$H = D \cdot q$	rem	Sv (Sievert)	$1\ \text{Sv} = 1\,\dfrac{J}{kg} = 100\ \text{rem}$

E   = energy
m   = mass
rad = radiation absorbed dose
C   = Coulomb

R   = Roentgen
rem = Roentgen equivalent
q   = evaluation factor for
        radioactive absorption
        of biologic bodies

Table 2.6. Threshold values of radiation doses for human organs

gonads, uterus, red bone marrow	0.3 mSv
bone surface, skin	1.8 mSv
all other organs and tissue	0.9 mSv

Special scintillation counters are able to measure the abundant energy of radiation by the strength of the sparkles, which is recorded in MeV. Most common are the channels 1.46 MeV, 1.76 MeV and 2.62 for the natural radionuclides ^{40}K, ^{238}U and ^{232}Th, which allow the differentiation of the three elements.

Radiometric surveys are carried out in the field in dense grids. The scintillometers are either held close above the surface of the earth (0.2–1.0 m), or placed in shallow percussion holes 0.5–1 m deep. Even airborne surveys are common from helicopters or airplanes, which fly very low. The results are presented mostly in colored isoline maps or/and in radiometric sections of the γ-intensity.

This method is used to locate buried radioactive waste and the γ-radiation that emerges from radioactive rain as fallout. But also geological rock complexes, which contain radioactive minerals like granit or uraniferous lodes, can be prospected. Radiometric results can tell about the spread of radioactive material on the surface or very near to the surface. Radionuclides that sit deeper than 3 m in the ground may already be out of reach. Some but not all domestic waste sites display weak γ-activity, which can be used for detection.

Radon surveys count the α-radiation that is emitted by the decay of the radon isotope. The datum is the content of radon gas of the ground air. A limited amount of ground air is pumped out of 0.5–1-m depth, after a hole has been sunk by percussion. The number of radioactive decay events is measured at the air sample by an α-scintillometer or an ionization chamber.

The emittance of radon by contaminated waste or by natural rock is based upon geochemical and geodynamical processes. Radon enrichment of the soil indicates the presence of preferred ascent paths. This can be leaking seals at dumps or natural steep dipping structures, like fracture or fissure zones. In any case, gases contaminated by radon will follow the path that is paved by higher permeabilities of rocks or materials. Radon surveys are therefore well adapted to investigate planned waste disposal sites in hard rock for potential transport paths of contaminants.

2.8 Isotope Hydrology

This method is based on isotopes, which may have a natural or geogenetic origin, and environmental isotopes, which are created by human activities. Some isotopes are radioactive; others remain stable. The first group consists of the isotopes of the radioactive carbon ^{14}C (half-life period 5730 years) and of radioactive hydrogen ^3H, called tritium (half-life period 12.43 Years, Fig. 2.29). The second group includes the isotopes of hydrogen ^1H and ^2H (deuterium), the isotopes of oxygen ^{16}O and ^{18}O and of carbon ^{13}C.

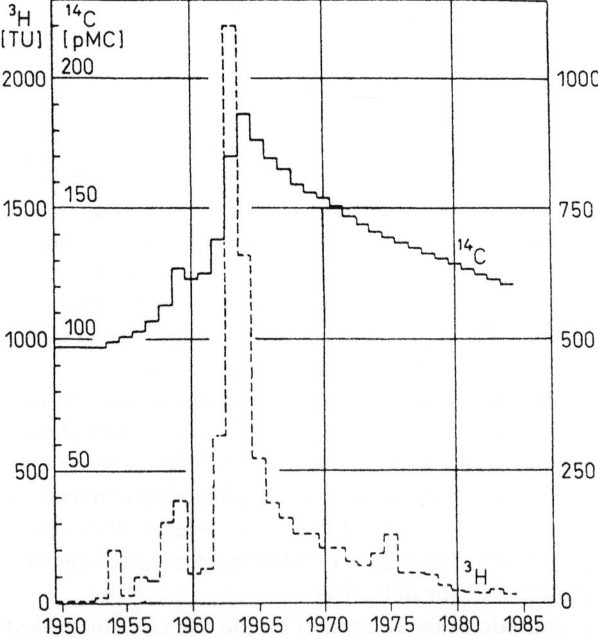

Fig. 2.29. Tritium (^3H) and radiocarbon (^{14}C) in CO^2 of European precipitation from 1950 to 1985

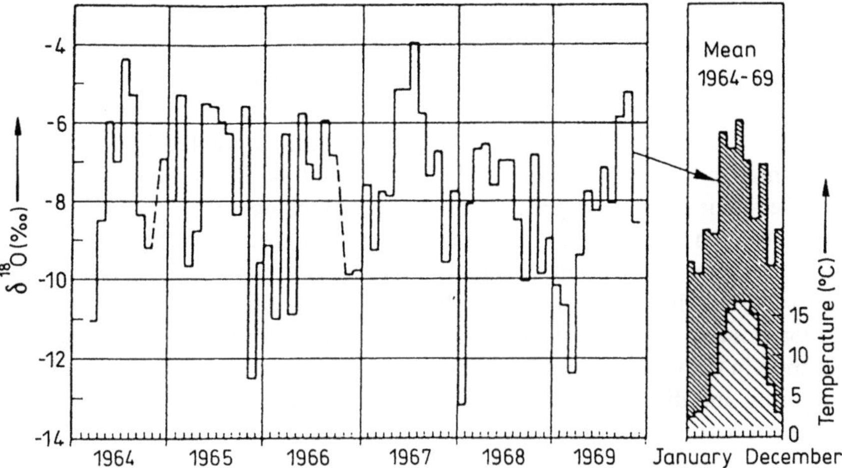

Fig. 2.30. σ-^{18}O content and mean yearly temperature in Groningen/Holland

Such isotopes are tools of the environmental geophysics that are used to estimate the amount of ground water regeneration and its dynamics, to differentiate ground water horizons by their genesis, and to estimate the share and the origin of toxic constituents in ground water. The determination of the portions of hydrogen and oxygen isotopes in ground water can be done fast and cheaply. The ^{18}O/^{16}O relation of a water sample is called its σ-value.

Stable isotopes are components of water molecules with different atomic weights: $^{1}H_2^{16}O$, $^{1}H_2H^{16}O$, $^{2}H_2^{16}O$ and $^{1}H_2^{18}O$. In these molecules, the atomic masses 18,19 and 20 are predominant. Their environmental importance stems from their different steam pressures, which lead to the enrichment of the lighter molecules in the volatile phase by evaporation, condensation or sublimation of water. This effect is called isotope fractioning.

The relation between hydrogen and oxygen isotopes is measured mostly not by its absolute value but comparatively, by its VSMOW-delta-value. It comes from Vienna Standard Mean Ocean Water σ-value, scaled in ‰ (parts per thousand). If ground water is accumulated by precipitation, it is called meteoric water. Its $\sigma^{18}O$ and $\sigma^{2}H$ rates have a linear relation, which is called MWL or meteoric water line. With open and stagnant water, this relation is no longer linear, since evaporation leads to isotope fractioning.

The heavier water, molecules evaporate faster at higher temperature. Therefore the isotope configuration of the precipitation shows a yearly variation, which is distinctly different for summer and winter. This effect can be observed only in ground water that is not older than 4 years. It is, for example, possible to assess whether contaminated ground water is older than 4 years (Fig. 2.30).

Moreover, this temperature dependence allows the distinction of ground water, that was formed during the cold Pliocene period from the more recent ground water of the warmer Holocene.

The reduction of air temperature with growing altitude produces lower delta values of the rain. This is the key to determining the topographic altitude of the catchment area of spring water with an accuracy of ± 50 m.

The age assignment of hundreds to thousands of years-old ground water is substantially based on the radiocarbon isotope ^{14}C, which is created by the omnipresent cosmic radiation. The radiocarbon is inhaled as the gas CO_2 by living beings. After their death, the process of decay creates new CO_2, which is dissolved as hydrogen carbonate in rainwater and later in ground water.

During the aging of ground water, its radioactivity decreases regularly, according to the half-life time of radiocarbon. This means that by measuring the radiocarbon content, the age of the ground water could be calculated. However, not all carbon in the water comes from the sky. Other carbon molecules stem from the weathering and dissolving of limestone. They are the culprits for the unwanted permanent hardness of ground water.

This extra carbon may lead to ages of ground water that can be several thousand years too high. Therefore these ages have to be corrected. The difference between corrected and non-corrected ages is constant and is used to identify ground waters of equal origin or to follow-up movements of special ground water.

Lesser water ages are dealt with by the tritium (^3H) method. This has been caused by nuclear testing in the atmosphere and has later seeped into the ground. The nuclear test program reached its peak during the years 1963/64 (Fig. 2.29). Today there is none of this tritium left and any age determinations can operate only by the qualitative analysis of ground water that has been formed ever since.

The contents of tritium in water are scaled in tritium units = TU. 1 TU corresponds to the very weak concentration of one tritium atom to 10^{18} molecules of water. In many countries, the International Atomic Agency of the United Nations in Vienna performs continuous ^3H monitoring to study the short-term ground water circulation and to control the adherence to the nuclear test ban treaty.

In the environment, especially in the exploration of hazardous waste sites, the isotope hydrology may be employed in many ways. Three examples illustrate this:

A) It can be decided whether the salinity of ground water originates from great depths, or from dumps of salt mines.
B) At old watered mines, it can be established when its water reserves are exhausted by pumping.
C) The age of contaminated leachate plumes can be determined.

3 Case Histories

3.1 Abandoned Hazardous Sites

3.1.1 Preconditions

Abandoned hazardous waste sites or old landfills may contain or emit substances that contaminate the ground, the ground water and the biosphere. Geophysical exploration is necessary for any risk assessment at hazardous sites and landfills.

An important factor of geophysical application is its industrial safety in fieldwork. The protection of working personnel is better than at investigations by mechanical penetration such as drilling, probing or trenching, since these may set free harmful gasses or liquids. In addition, geophysics is not destructive, is non-invasive and can cover the total lateral and spatial expanse of a waste site, whereas boreholes can only inform about one point or a thin column.

Nevertheless, geophysical data should never stand alone. They should always be supplemented and verified by limited drilling or probing. This combination furthermore allows the direct calibration of a geophysical method to the special conditions of a certain hazardous site and the preparation of appropriate remedial techniques.

3.1.2 Geomagnetics

Geomagnetic Survey

Geomagnetic measurements by proton magnetometer, to determine the total magnetic intensity, should be applied in most cases, since it is a cheap, fast-progressing and simple field method. The following case history will prove this:

Figure 3.1 shows anomalies of the magnetic total intensity, which were surveyed in a rectangular grid of 2×2 m over a buried hazardous dump, known to contain industrial sludges, besides domestic waste. The result is plotted as a 3D picture and one can distinguish two categories:

1. Extremely high singular anomalies > 1000 nT,
2. Weak anomalies 20 to 150 nT.

The high values are related to ferrimagnetic bodies, which lie less than 1-m deep. The weak anomalies originate either from similar small bodies, which are buried up to 3 m deep, or from bigger, but less magnetic bodies. Such accumulation of

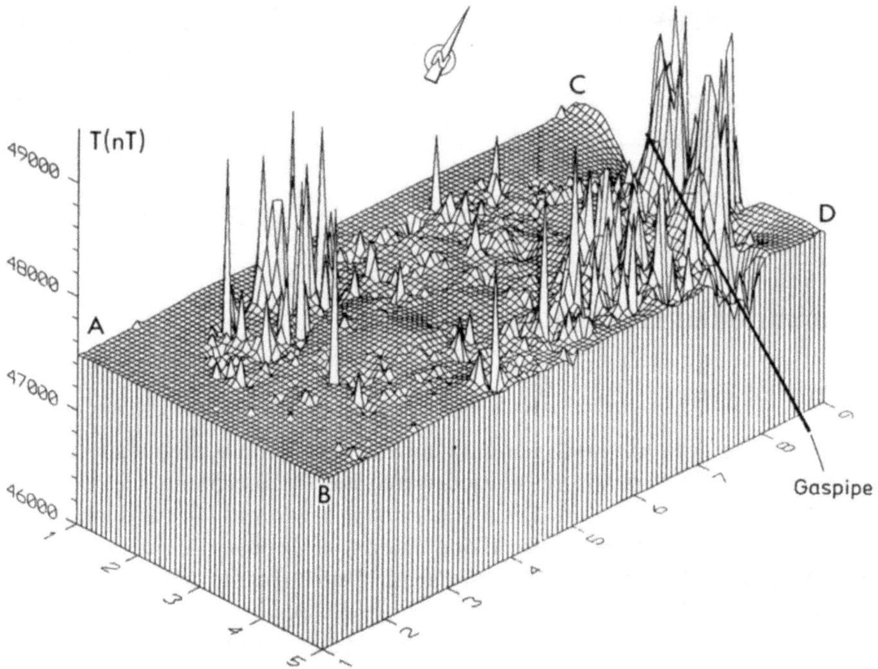

Fig. 3.1. 3D picture of the magnetic total intensity of a hazardous waste deposit containing domestic and industrial waste

small, point-related magnetic anomalies is characteristic for most domestic waste deposits. Their magnetic sources are manifold, consisting of tins, wire netting, scrap of cars, etc., distributed at random throughout the dumped material.

The wider magnetic anomalies of Fig. 3.1 depict local assemblages of scrap which should be situated in the southwest and in the northeast. In the same area, a gas pipeline, made of steel, follows a forest trail and enhances the scrap anomaly.

Over the same deposit, the vertical gradient of the total intensity was calculated (Fig. 3.2 and Sect. 2.1.1) from two measurements, 0.65 and 1.65 m above ground. Two protonmagnetometers were attached to a non-ferrous rod at these heights and were simultaneously recorded. Compared with the simple measurement of the total intensity, the gradient is less influenced by the small variations of the magnetic field, which are known as "magnetic noise." In areas of small magnetic anomalies, it enhances those anomalies that originate from ferrimagnetic material.

The border of the covered-up waste deposit can be derived from 3D presentations of both the total intensity and the gradient survey (Figs. 3.1 and 3.2) by connecting all marginal magnetic anomalies.

Figure 3.3 shows the result of a geomagnetic survey of old gasworks. At this abandoned industrial site, the geomagnetic survey met very difficult conditions

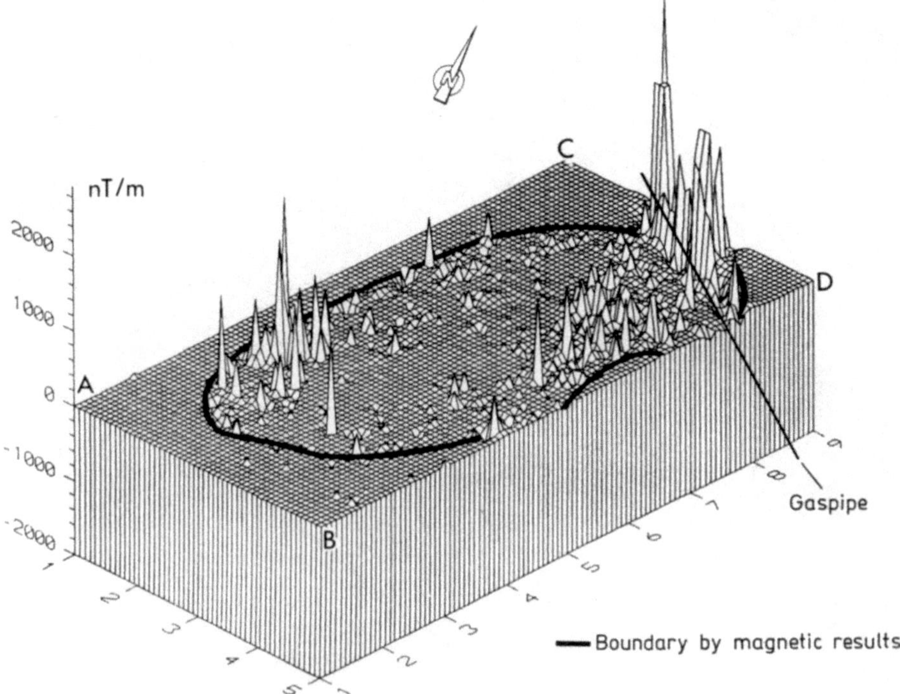

Fig. 3.2. 3D picture of vertical gradient of magnetic total intensity (hazardous waste deposit Fig. 3.1)

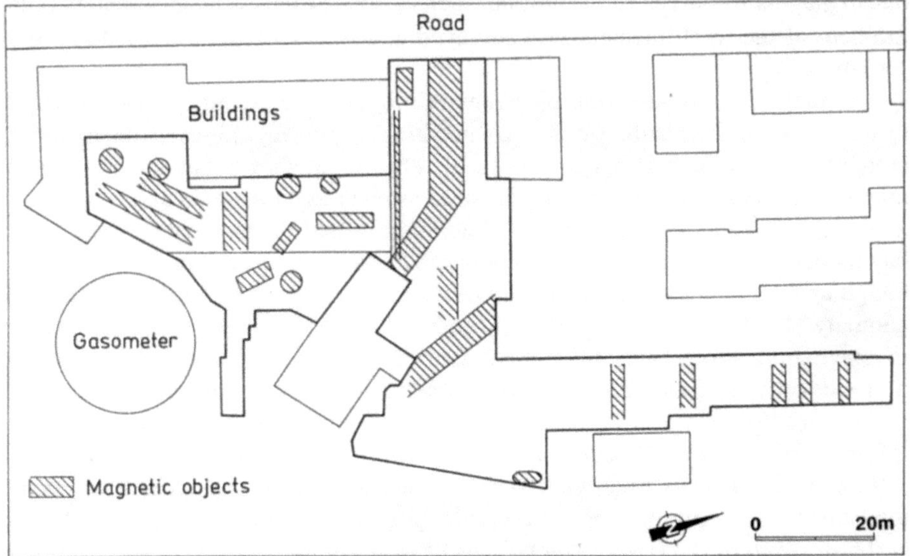

Fig. 3.3. Magnetic map of old gasworks within built-up area

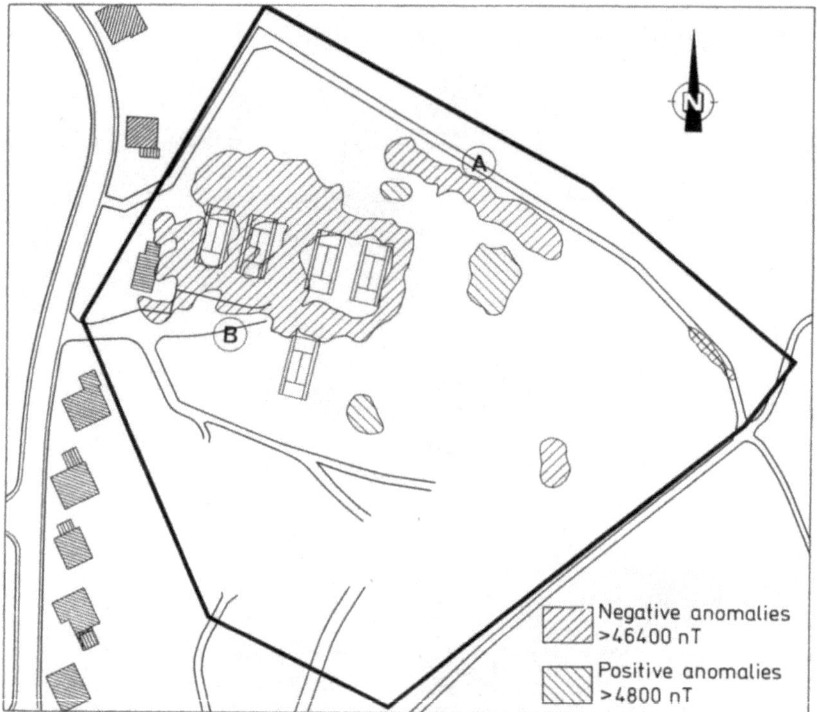

Fig. 3.4. Anomalies of magnetic total intensity over former landfill, now used as tennis club

due to magnetic sources above ground, such as houses, tanks and gasometers. The gradient of the total intensity was surveyed at 1.0 and 2.0 m above ground in a 1×1-m grid.

The magnetic investigation pinpoints reinforced fundaments, steel pipes and iron troughs in the shallow underground, in spite of the magnetic disturbances created by the industrial buildings nearby. This result shows that narrow-spaced magnetic surveys can be applied with success even in built-up areas.

Another magnetic target was an abandoned waste disposal site which, after having been covered up with soil, is now used as a tennis club (Fig. 3.4). Here the total magnetic intensity was measured in a quadratic pattern of 4 × 4 m. Under the anomaly "A" in the northeast, illegally dumped car wrecks were found. The other magnetic indications have still to be unearthed. They are a mixture of the magnetic fields of undergound bodies and of the wire net fences of the tennis courts.

This example makes clear that magnetic interpretations have to consider the influence of magnetic structures below and also above ground.

Figure 3.5 displays four steel barrels which have been put upright into non-magnetic silt. The proton magnetometer recorded a concentric anomaly of 110 nT at 0.5-m depth of the cover, which fades to 40 nT at 2.0-m overburden. The iso-line interval is 10 nT.

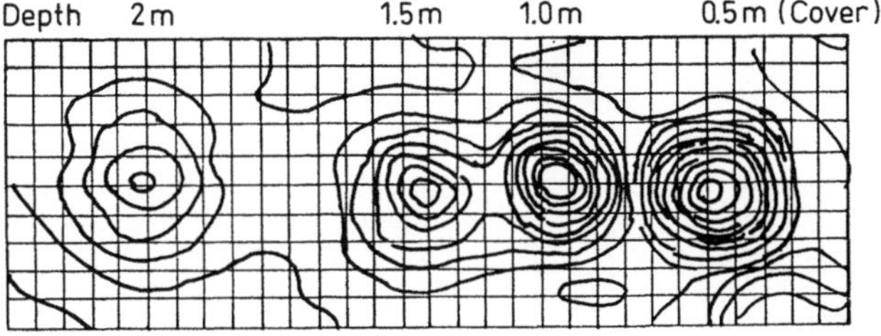

Fig. 3.5. Magnetic total intensity over four steel barrels, buried upright at different depths

This shows that single barrels or bundles of drums produce only weak magnetic anomalies, which are often smaller than the common magnetic variations of domestic or industrial waste. In addition, their magnetic response decreases rapidly with depth. Barrels, perhaps with toxic fillings, can be located only when they are embedded in non-magnetic material.

On the other hand, Fig. 3.5 shows that a 1×1-m rectangular grid suffices to find separate anomalies of singular magnetic objects, provided their magnetic field is stronger than the background.

Other test measurements have revealed that single metal barrels may be magnetically detectable down to a depth of 4 m and bundles of between 20 and 100 barrels down to 10 m. However, the precondition must always be fulfilled that no other magnetic materials have been buried nearby.

The next case history concerns a subterranean air-raid shelter that was known to house metal containers filled with contaminant liquids. The outcome of the geomagnetic investigation is shown in Figs. 3.6 and 3.7 as a magnetic map and 3D picture. The total intensity shows a negative anomaly in the north and a positive in the south, as to be expected by a big magnetic object at a latitude of approx. 55° north within an inclined magnetic earth field.

An air-raid shelter lies in between the two anomalies. The peak reaches >2000 nT and complies totally with the magnetic field of the steel reinforcements of the concrete roofs and walls. There is no trace of the expected containers. A separate magnetic anomaly in the southwestern corner of the area must belong to another buried magnetic object.

This abandoned air-raid shelter, whose position was only vaguely known, lay under a dense forest. The survey lines had to be cut through dense scrub at distances of 2 m to allow for the topographical and geomagnetical surveys. Such additional preparatory activities are expensive and should be known before planning or even bidding for a survey. The same is valid for vegetation or agricultural crops and topographical features like steep rocks or ravines.

The efficiency of a magnetic survey over a landfill in Indiana, USA, was considerably increased by calculating the upward continuation of the magnetic field. The observed magnetic data (Fig. 3.8a) display that off the landfill, anomalies

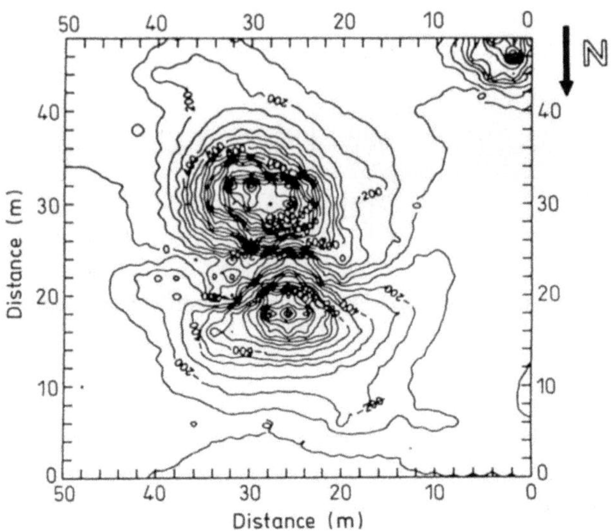

Fig. 3.6. Contours of total magnetic intensity of buried air-raid shelter with reinforced concrete roof and walls

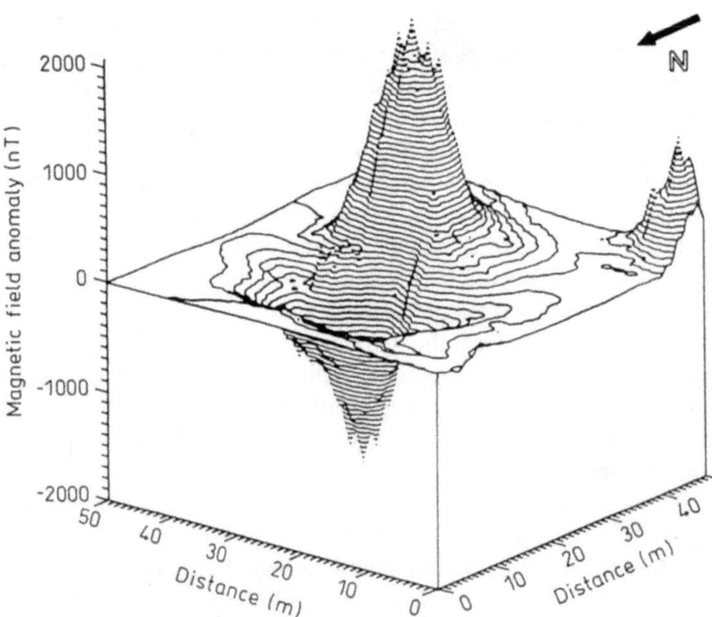

Fig. 3.7. 3D picture of magnetic total intensity, air-raid shelter Fig. 3.6

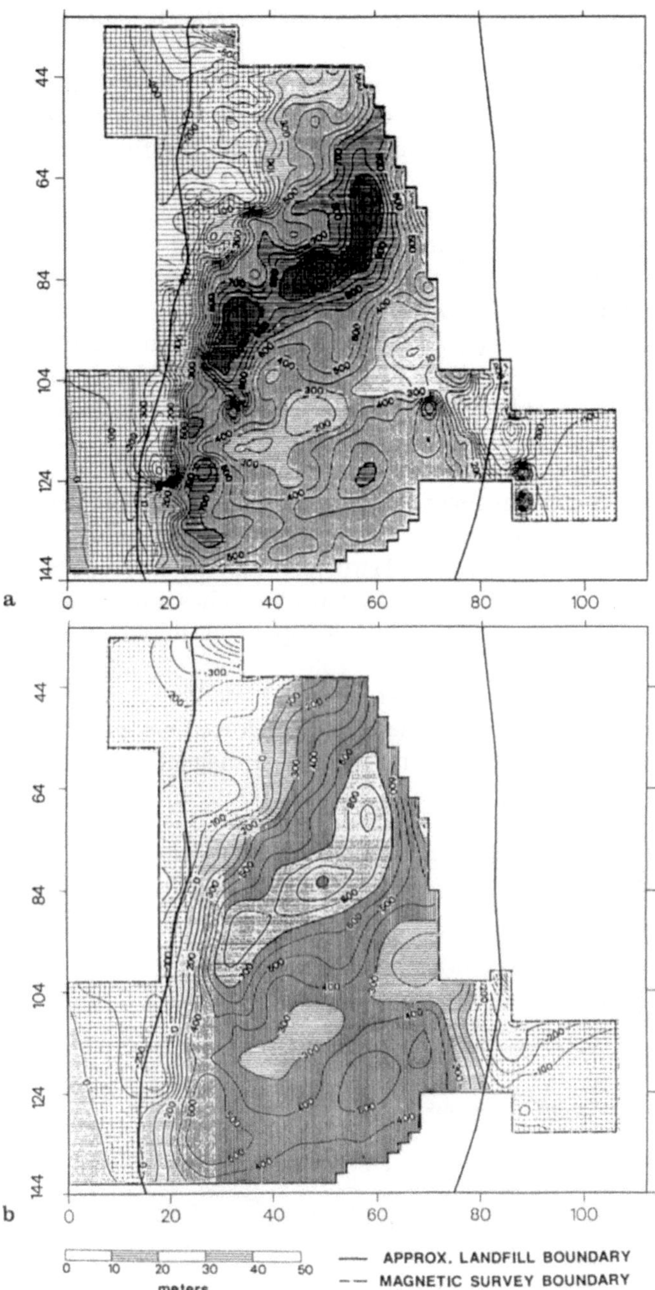

Fig. 3.8. a) Contours of magnetic total intensity at 1 m height, b) Contours continued at 6 m upward, landfill Indiana, USA

remain beyond 100 nT, except for anomalies by iron surface objects in the southeast.

The northwestern slope of the landfill is marked by local anomalies from small, shallow sources, appearing within an intense magnetic minimum. This is the negative counterpart to a strong, wide maximum over the landfill. The total amplitudes of both reach over 1000 nT.

In Fig. 3.8b, the magnetic field has been continued upward by calculation to a height of 6 m. This increases the distance of observation level to source and allows the attenuation of the distorting magnetic fields of the small objects. The landfill can now be viewed as a single magnetic source.

It is clearly demonstrated that simply by calculating upward continuations of the magnetic field, the center and the shape of a magnetic body, which is strewn with small intensive anomalies, can be clarified.

Airborne Magnetics

A magnetic survey by helicopter was flown in a height of 30 m over a big deposit of domestic garbage. Figure 3.9 proves that this kind of waste possesses ferrimagnetic properties and stands out as a special structure from the magnetic background. In the isoline map (Fig. 3.9), the total intensity reaches a maximum of >750 nT in the south of the waste dump. The corresponding minimum in the north sinks down to <−560 nT.

The zero-line between them extends east-west and marks approximately the middle of the dump heap. As in the case of the reinforced air-raid shelter, this general structure is caused by the normal inductive magnetization of the

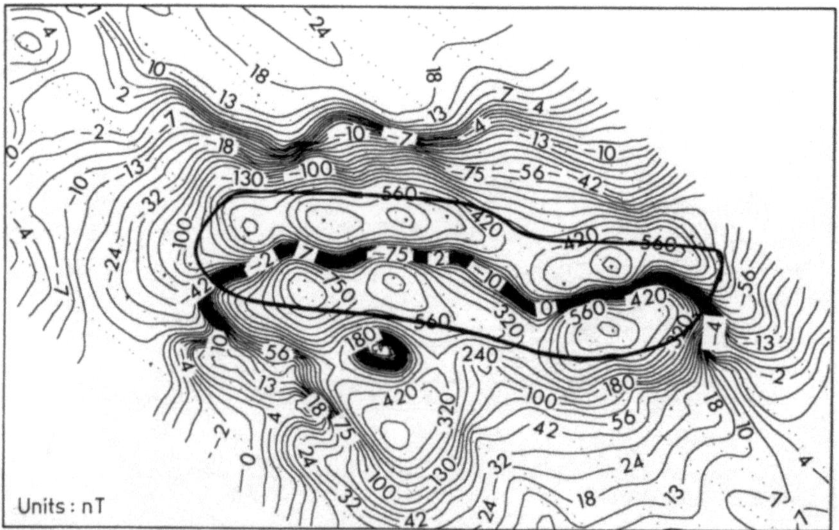

Fig. 3.9. Contours of magnetic total intensity by helicopter survey, piled-up domestic waste dump, Hannover, Germany

magnetic field of the earth. But there are also separate anomalies, especially in the southwest, which may pertain to concentrations of scrap iron that were dumped in the vicinity of the main deposit.

In spite of the low flying height, only a few separate anomalies were recorded. From this we must conclude that airborne magnetic surveys lack the resolution necessary to find single magnetic bodies or concentrations in a waste site. This task must be left to the more cumbersome ground surveys.

3.1.3 Geoelectric DC Methods

Like the magnetic, geoelectric methods are well adapted for sensing buried waste and waste migration. They work fast, at low cost and with high efficiency, since they not only can find the extension of a buried waste body, but also the location of singular contaminations of special resistivity, even at a depth exceeding 10 m. Such contaminations may be scrap metal, toxic sludges, organics of higher resistivity, or materials with low resistivity.

Geoelectric Mapping

Figure 3.10 shows the mapping of a domestic waste site by the Wenner array (see Sect. 2.2.1). By using an electrode separation of 10 m, very low specific resistivities between 13 and 30 Ωm have been surveyed over the dump. The lowest values occurred in the center; towards the edges, the values increase continuously.

It is surprising that such a homogenous distribution of low resistivity was found, in spite of the great variety of the many components of the domestic waste. Perhaps the single, electrically active components of the garbage are too small to create individual anomalies. Their geoelectric fields are obviously integrated into an overall homogeneous anomaly of low resistivity. Nearly all domestic waste sites display such a homogenous electric behavior and can be detected by this overall low resistivitiy.

It is assumed that not only the solid components lower the resistivity of domestic waste, but that leachates and seepages contribute as well. The general reason for this is thought to be a general salt content of domestic waste. The salinity of plumes, originating from waste sites, can furthermore be used as a geoelectric tracer. Such plumes, often loaded with toxic freight, can be followed up by geoelectric mapping.

The next case is a covered-up site of mixed domestic and industrial waste. It was mapped by a Wenner array, using an electrode distance of 5 m and a line separation of 7.5 m (Fig. 3.11). In agreement with the previous case, the waste is characterized by a general low resistivity $<20\ \Omega$m. The uncontaminated vicinity, however, has resistivities $>60\ \Omega$m. This pronounced difference allows the construction of the border of the dump. In addition, hints about how seepage channels may run are derived from the geoelectric map of Fig. 3.11.

Whereas the Wenner array is frequently employed for DC mapping because it is symmetric and the resistivity data pertain to the center of the array, other arrays

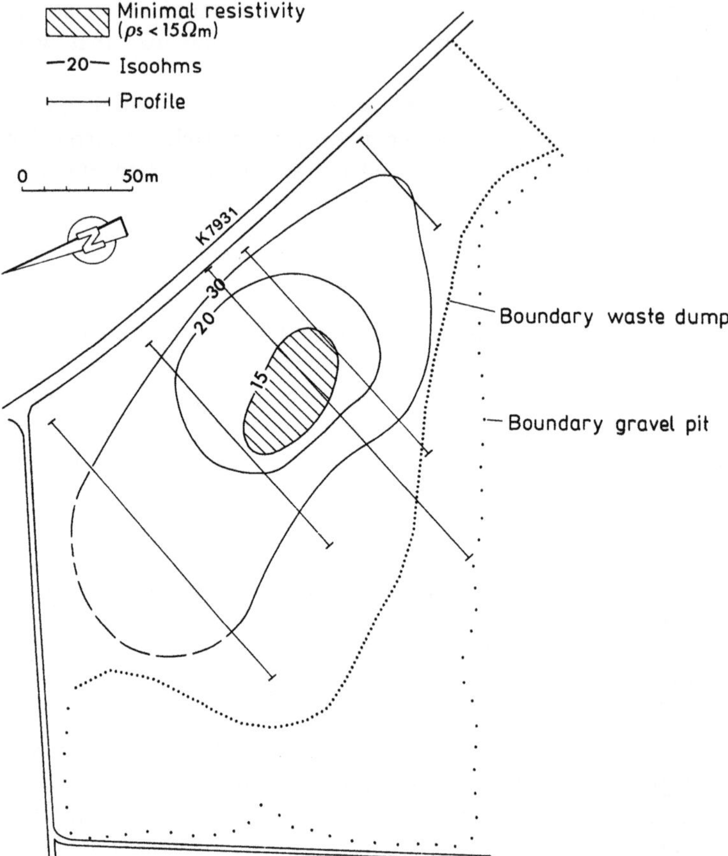

Fig. 3.10. Contours of apparent specific resisitivity ρ_s, domestic waste dump in the alpine foreland

may be better suited for special tasks. The gradient array has better depth penetration and better resolution of structural details. It should be chosen when additional information on the structural set-up and the geological barrier below waste
sites is requested. The fixed Schlumberger array has the advantage of gathering
specified information from a certain depth level, as shown in Fig. 3.12.

In the Panoche fan area of California ground water at the edges of the fan is
contaminated by Selenium. Figure 3.12 displays the contours of interpreted resistivity at a depth of 20 m. It was constructed from 82 Schlumberger DC
soundings as a geoelectric map of a fixed Schlumberger array.

The fan was found to be composed of coarse-grained sediments with specific
resistivities $>20\,\Omega$m (dotted in Fig. 3.12), while the surrounding clays are
characterized by specific resistivities $<7\,\Omega$m (stippled in Fig. 3.12). The selenium contamination is to be expected in the white area in between.

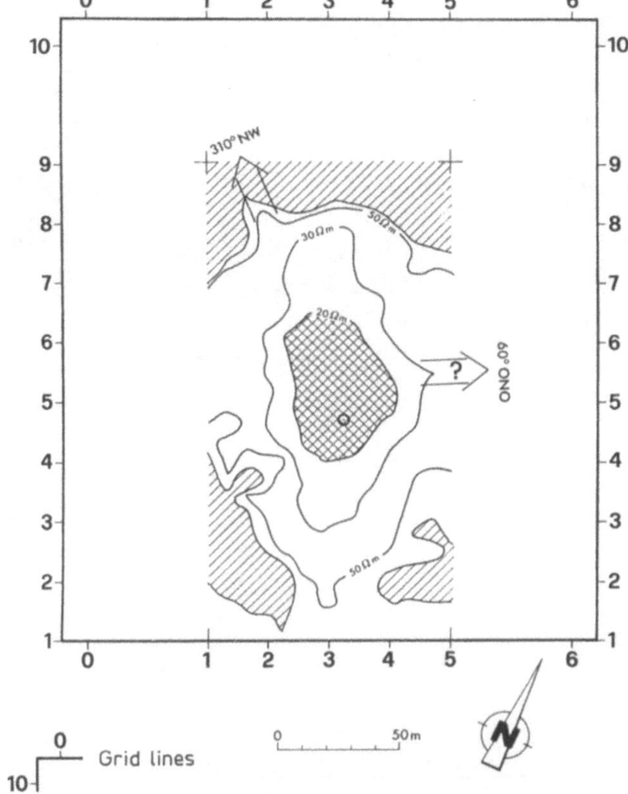

Fig. 3.11. Contours of apparent specific resistivity ρ_s, industrial/domestic waste dump. Area hatched > 60 Ωm, cross hatched < 20 Ωm. arrows = possible seepage paths

Vertical Electric Sounding (VES)

This method is the vertical complement of the lateral-directed geoelectric mapping. It has already been mentioned that the "Schlumberger array" mostly is used. This array results in the construction of a column under its midpoint. As in a cored drill hole, the boundaries or thickness of horizontal beds are recorded, but instead of petrography, the apparent specific resistivities of layers are presented.

Over a hazardous waste deposit, situated on an island in the river Rhine, 10 VES with a maximum electrode distance of 260 m, were made in a section. Figure 3.13 shows, at the top, two soundings with their results, obtained after mathematical inversion. All data are correlated to the geoelectric section at the bottom. The shapes of the sounding curves inside and off the waste site are completely different.

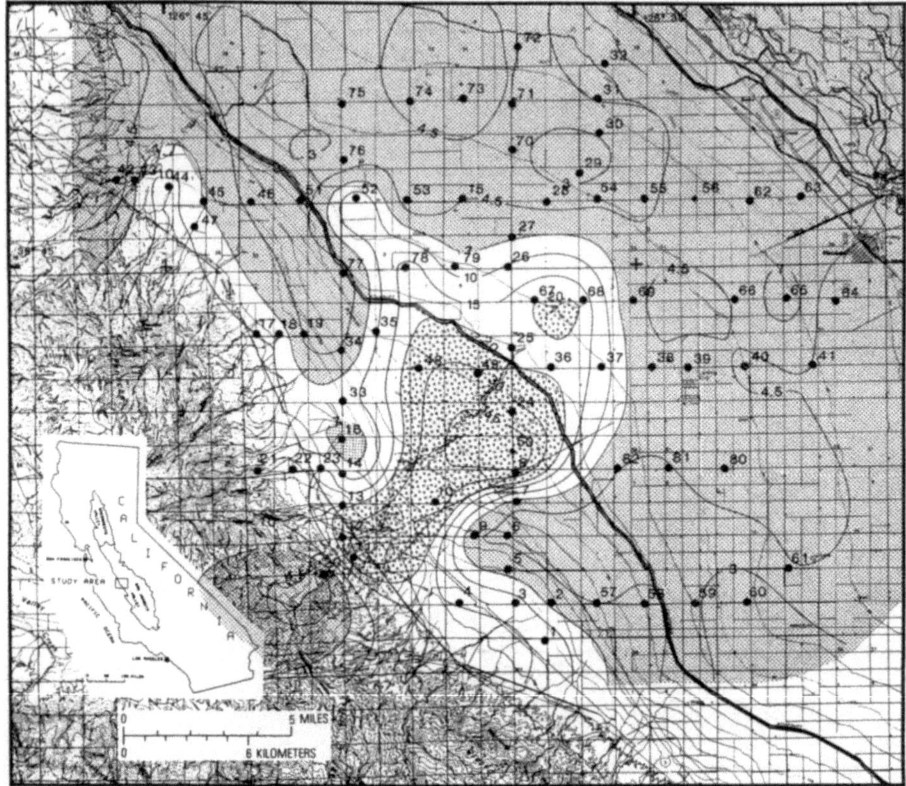

Fig 3.12. Contours of resistivity (Schlumberger soundings) 20-m depth, Panoche Fan, California. Resistivities $> 20\,\Omega$m are dotted, $< 7\,\Omega$m are stippled, white areas are contaminated by selenium

The curve 17 E in Fig. 3.13 was measured outside the dump. It has a strong maximum, which is created by a bed of wet sand and gravel, ~ 10 m thick with $650\,\Omega$m. It is covered by soil of 0.8-m thickness and a resistivity of $90\,\Omega$m. The bed below this aquifer reaches 48 m deeper and must be clay or clayish silt, since its resistivity drops to $48\,\Omega$m.

Inside the dump, the soundings resulted in minimum curves: The VES 13E found a covering layer of $94\,\Omega$m. Underneath, two beds of 8 and $23\,\Omega$m were assigned to the waste and are together ~ 20 m thick. The strata below again reaches a resistivity of $\sim 90\,\Omega$m.

The interpreted geoelectric section at the bottom of Fig. 3.13 shows that the same difference in curve shape exists for all curves, surveyed inside or around the waste site. In spite of wider variations of resistivities inside the dump, the premise of homogenous and low resistivity for waste is again fulfilled.

Figure 3.13 is a good example for the application of geoelectric soundings to waste sites. Not only the resistivity of the deposited material can be determined,

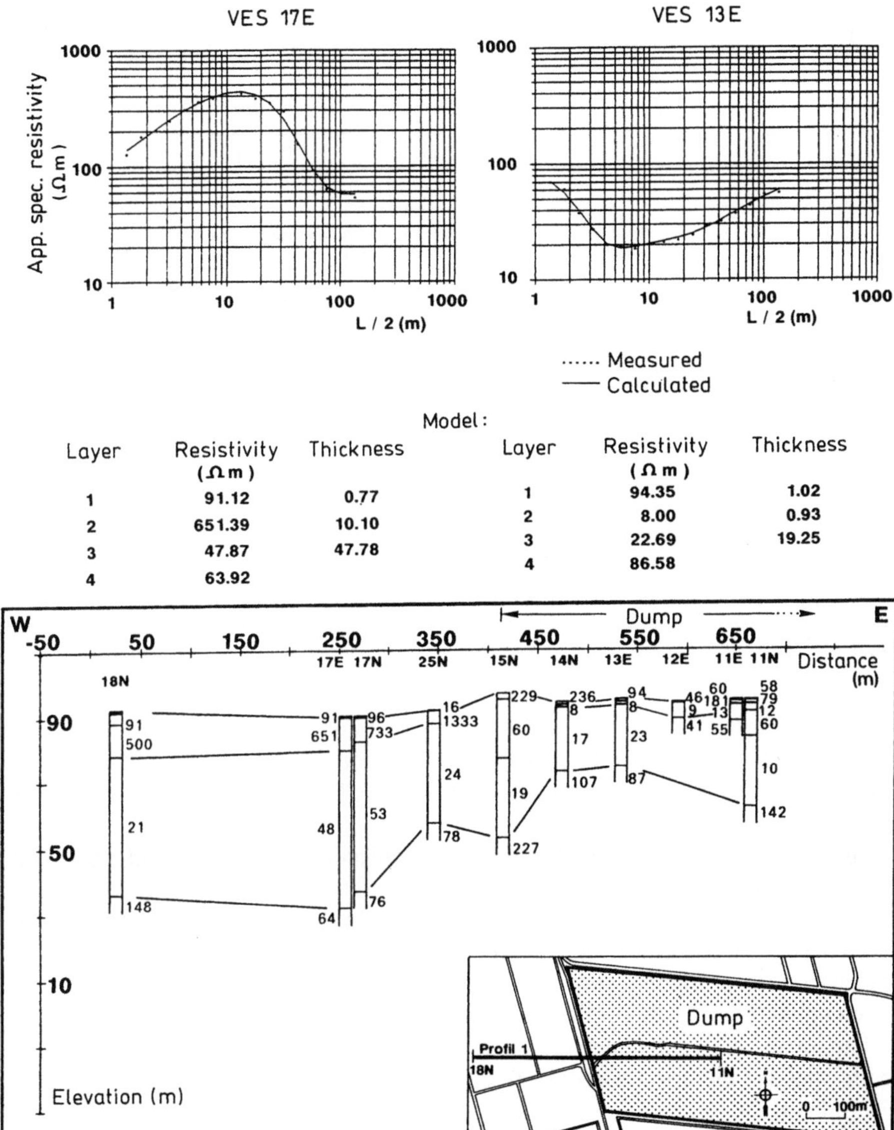

Fig. 3.13. Geoelectric soundings (Schlumberger array) of hazardous waste site on Rhine island. Sounding 17E lies outside, sounding 13E inside the dump, displaying profound differences in resistivity between waste and sediment

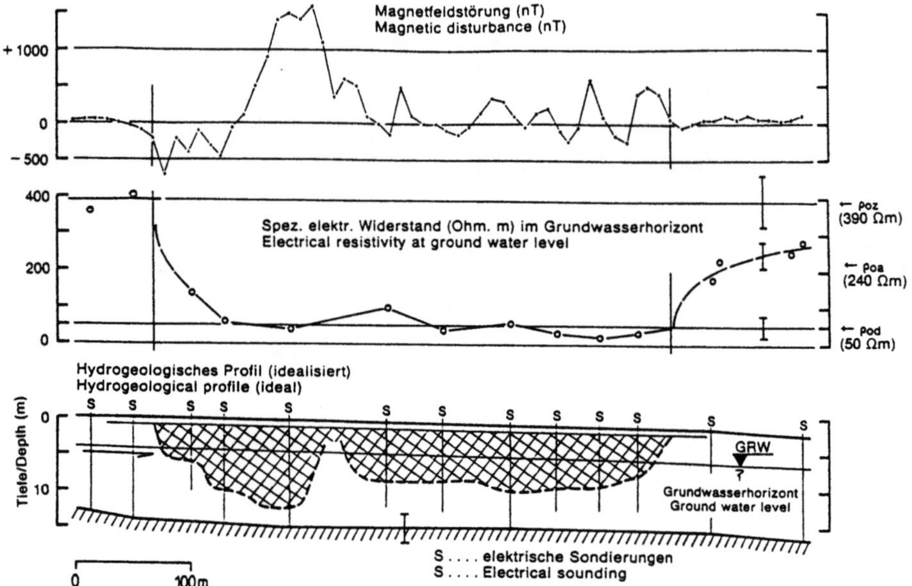

Fig. 3.14. Combined investigation of contaminated ground water by geomagnetic measurements and VES

but also the thickness of the base and the extension of the dump can be worked out. Soundings are helpful to pinpoint the location of special waste components. In this case, the resistivities around 200 Ωm come from casting sands, which were dumped many years ago.

Figure 3.14 describes the detailed investigation of a saline ground water contamination, which stems from a recultivated hazardous waste dump. At first, the rims of the site were located by geomagnetic mapping (upper section). Then the thickness of the dump was delineated by electric soundings (VES) (lower, cross-hatched section). Finally, the average specific resistivities at ground water level were connected (middle section).

The intrusion of salty seepage water into the aquifer results in a rapid decrease in the specific electric resistivity from >300 Ωm to <50 Ωm over a distance of 130 m from the left rim.

The ground water flows from the left to the right of the picture. The water resisitivity remains low over the whole extension of the waste deposit of 800 m. At its right end, the resistivity of the downwash plume increases again and reaches 280 Ωm at a distance of 250 m.

It is surprising that the concentration of soluble salts and perhaps of soluble contaminants diminishes to 15% of its maximum value over the distance of only 250 m.

VES surveys can be applied not only to single plumes but also to regional distributions of briny ground water. This requires much greater depth penetration reached by increased electrode separations up to 10000 m.

Figures 3.15 to 3.17 present VES results from the Upper Rhine Valley between the extinct volcano Kaiserstuhl (Whyl) in the south and the town of Strassbourg in the north. Over a distance of 60 km, 11 soundings with a maximum electrode distance of 4000 m were made. By this, a depth of >1300 m below the surface of the valley was reached.

The upper section of Fig. 3.15 shows the geology and block tectonics of the tertiary and mesozoic strata, constructed by surface outcrops and drilling results. This is confronted by the geoelectric section. In contrast to the multiple structures of geology, it can be divided into three areas only:

1. Upper zone, down to 300 m. Within the total length of the section, varying resistivities up to 450 Ωm prevail. They are attributed to gravel beds of quaternary age.
2. Second zone. Its resistivities range from 20 to 80 Ωm. In the south, they start at a depth of 150 m, in the middle they sink at extension faults to 400 m, and in the northern part they begin at a depth of approximately 1500 m. They are assigned to the mesozoic series.
3. Northern zone. The lowest resistivities from 2.5 to 9 Ωm occur only in this part. They extend to a depth of >1000 m, above the downthrusted mesozoic strata. They are correlated to beds of tertiary age, which normally have specific resistivities >100 Ωm.

The reason for this was detected by a drill hole at the village of Marlen near the town Kehl, which lies at the north end of the section. It is subterranean ground- and pore water of high salinity, which originates from naturally leached salt deposits at the base of the tertiary series.

This geoelectric discovery of the great depth extension of saline ground water, south of Strassbourg (Fig. 3.15) has proven that this contamination is of natural origin and is not caused by old stocks of abandoned salt mines.

The VES 08 is located near the borehole of Kehl-Marlen. In Fig. 3.16, the thickness of beds and their specific resistivities are shown for the soundings 08 and 09. At VES 08, the resistivities vary between 239 and 67 Ωm, down to 215 m. Below this they sink to only 9 Ωm. This agrees very well with the result of the borehole that was sunk at 215 m into saline ground water.

The maximum electrode separation of VES 08 was 1600 m, but even by this considerable spacing the bottom of the saline water or the boundary between tertiary and mesozoic strata was not reached. Only when the electrode spacing was prolonged to 3800 m in the nearby VES 09 was the bottom of salinity finally found at 1500 m. At this depth, the steep decline of the sounding curve to the extremely low resistivity of 2.5 Ωm changes over into a minimum and an increase to 20 Ωm, thereby marking the top of the mesozoic impermeable series (see middle beam of Fig. 3.16).

A strong contrast is displayed by the curve of VES 04 in Fig. 3.17. It lies in the southern part of the section. The lowest resistivities are 30 Ωm and belong to mesozoic rocks. They extend from the depth of 163 to 1137 m. The quaternary and latest tertiary beds (Pliocene) show resistivities from 190 to 270 Ωm. They

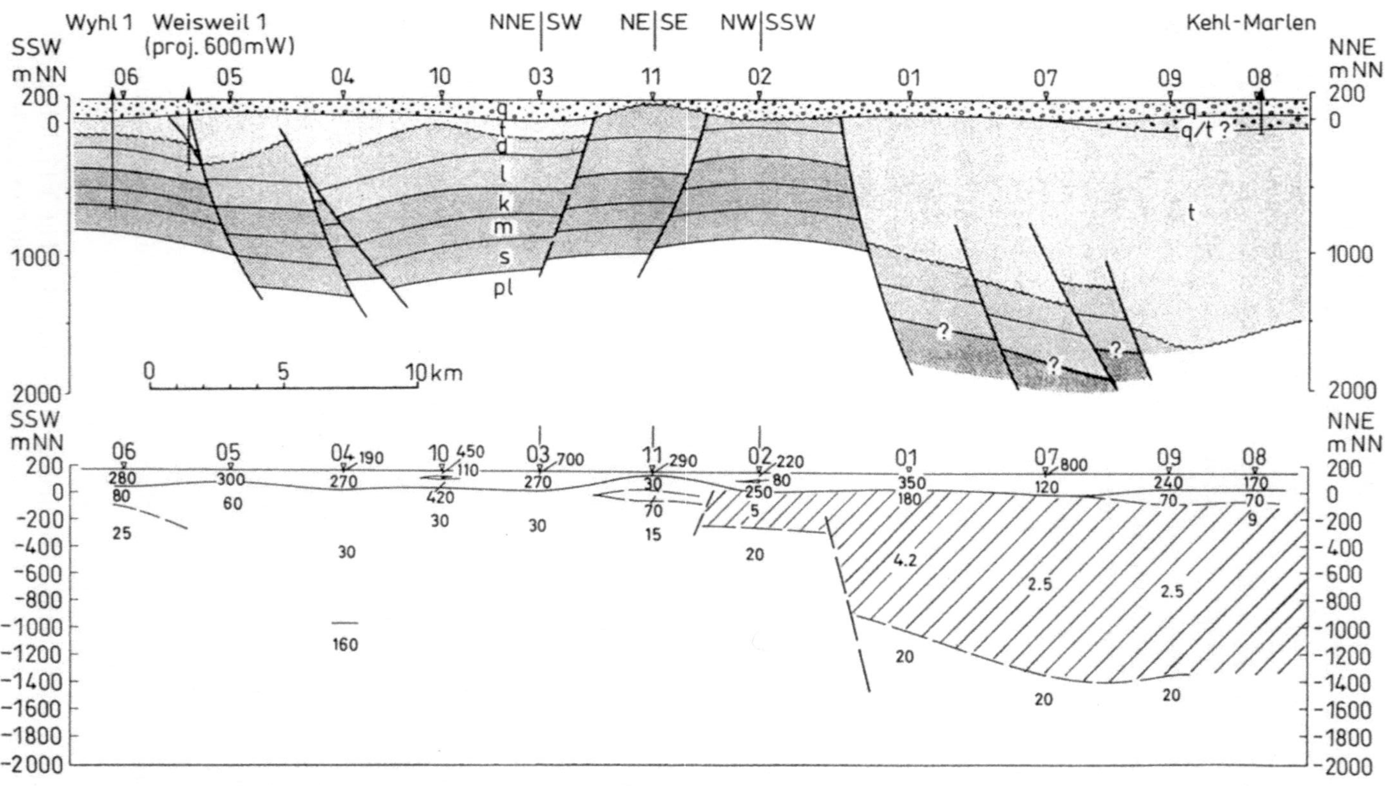

Fig. 3.15. Geology and geoelectrical profile of the Rhine valley from the extinct volcano Kaiserstuhl to Strassbourg

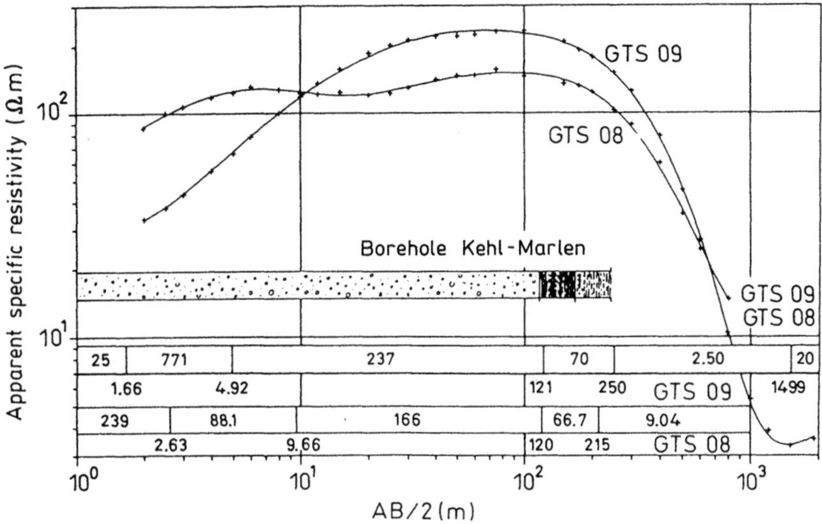

Fig. 3.16. VES curves 09 and 08 in north of Fig. 3.15, near Strassbourg

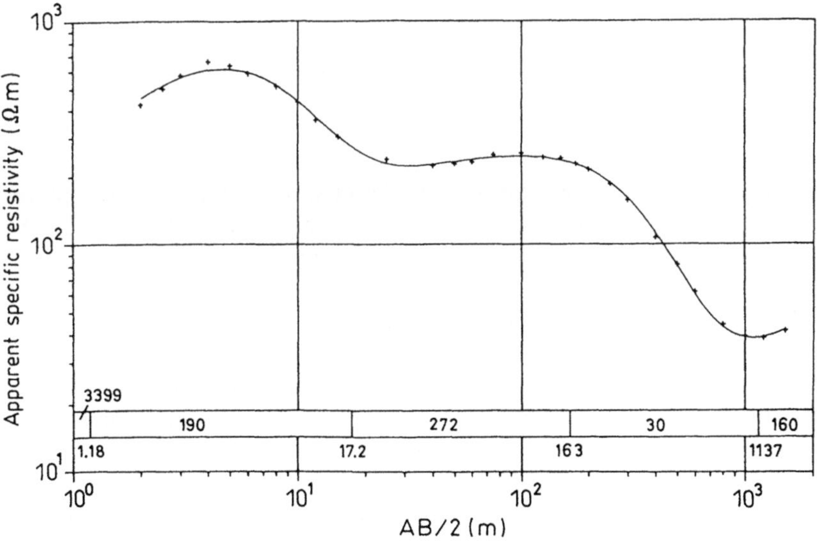

Fig. 3.17. VES curve 04 in south of Fig. 3.15, near Kaiserstuhl

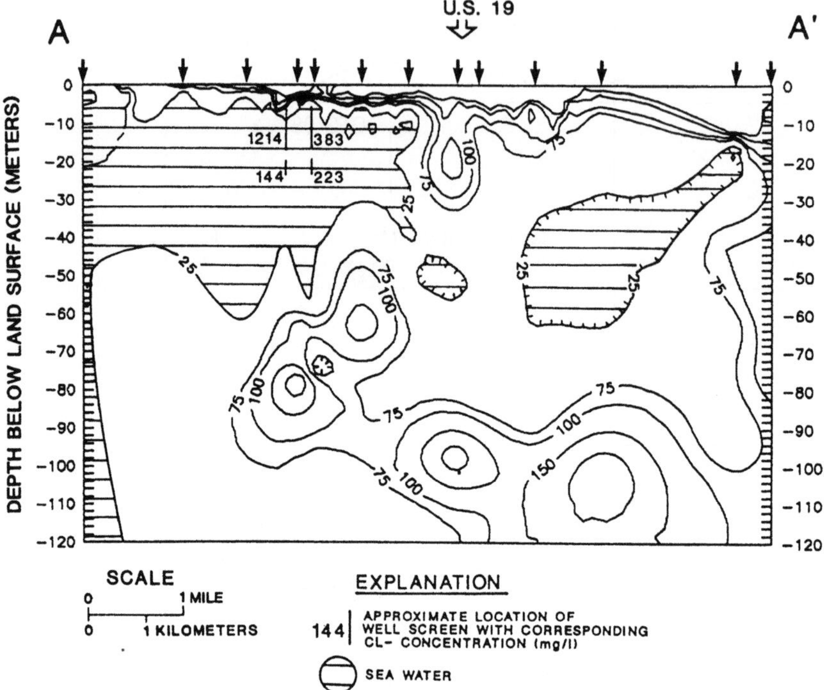

Fig. 3.18. Geoelectric profile north of Cross-Florida Barge Canal. Well data are included, arrows mark VES

are underlaid by a thick succession of stratified rocks, which are correlated to tertiary and mesozoic strata, as displayed in the geological section of Fig. 3.15. The reascent of resistivity at the depth of 1137 m is probably caused by basement rocks. But there is no trace of the very low resistivities $<5\,\Omega$m, indicating the high salinity of ground water as in the north of the section.

The spreading of this gigantic plume was controlled by repeating these 11 soundings two years later. The result was a surprise, since no change or extension could be found. This confirmed the assumption of a deep-reaching salinization as a remnant of intensive exhaustion of a tertiary salt deposit.

The investigation of a salt-water intrusion into an aquifer by direct current resistivity soundings (VES) in Florida is presented in Figs. 3.18 and 3.19. A sea-level canal is cut 4 m into karstic carbonate rocks comprising an important aquifer. This causes the intrusion of saline water from above and the upwelling of mineralized sulfate water from below (Fig. 3.19).

The sea-water seepage under the canal is outlined by the 25 Ωm contour. Its total area of mixing and encroachment is narrow, but extends to a depth of 40 m. It is characterized by resistivities $<75\,\Omega$m. At a greater depth under the river, a well detected that the 75 Ωm contour encloses ground water with low chloride but high sulfate concentration (~ 330 mg/l).

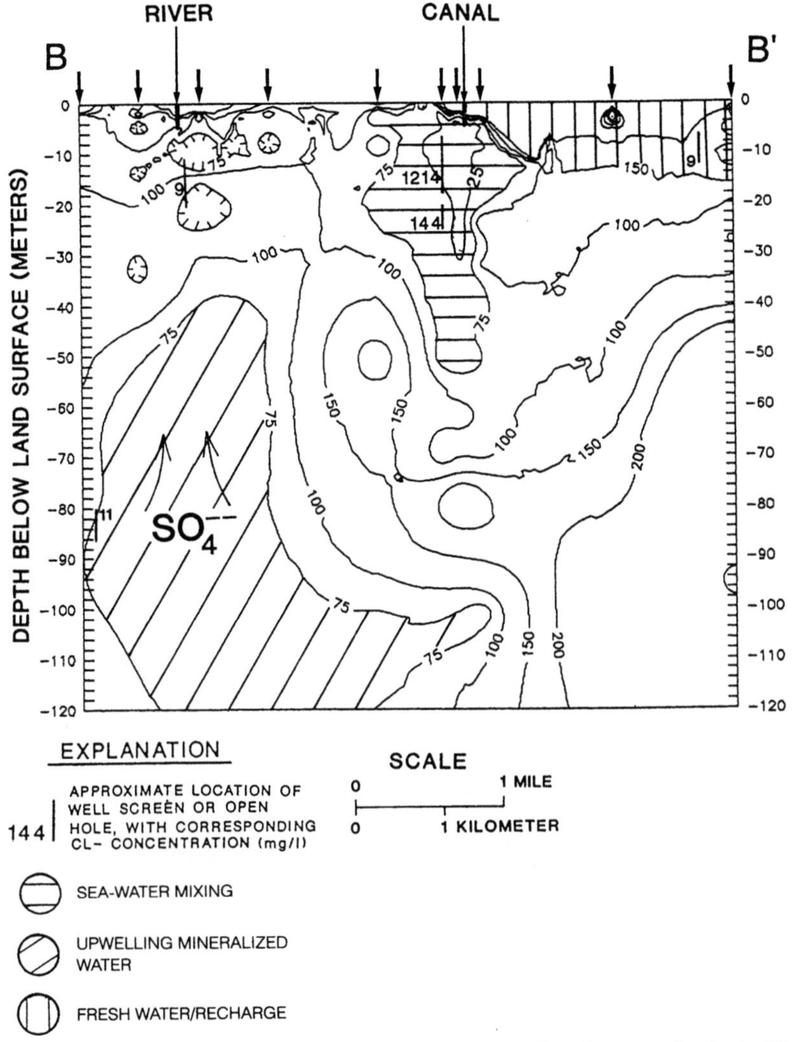

Fig. 3.19. Geoelectric profile of Cross-Florida Barge Canal, perpendicular to Fig. 3.18

A dipole-dipole resistivity survey at two chemical disposal ditches in northern Utah was very successful, despite its complicated structural background. Figure 3.20 shows the result at this disposal site of solvents. The profile was measured by a single seven-electrode spread with a dipole length of 9.1 m.

While the upper section of observed data shows low resistivities from below the ditches to the lowest level of measurement, the numerical model clearly shows that resistivities $<10\,\Omega$m occur separately within a small area directly under the ditches and again as a thick layer at greater depths.

From this model and the section of the computed apparent resistivity, a geologic interpretation is derived. The complex structures consist of layers of del-

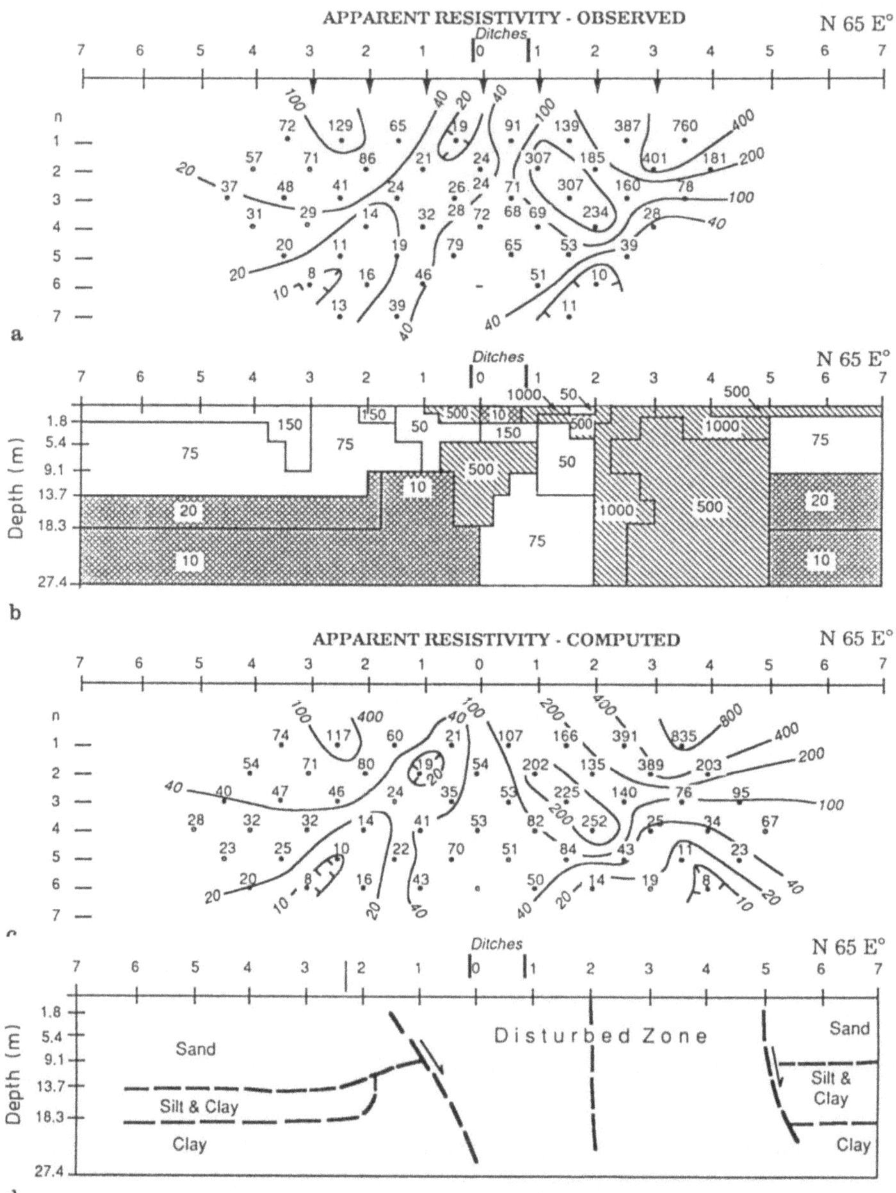

Fig. 3.20. Resistivity profile of chemical disposal ditches, northern Utah, dipole-dipole array. Lake-bed sediments, disrupted by slump block

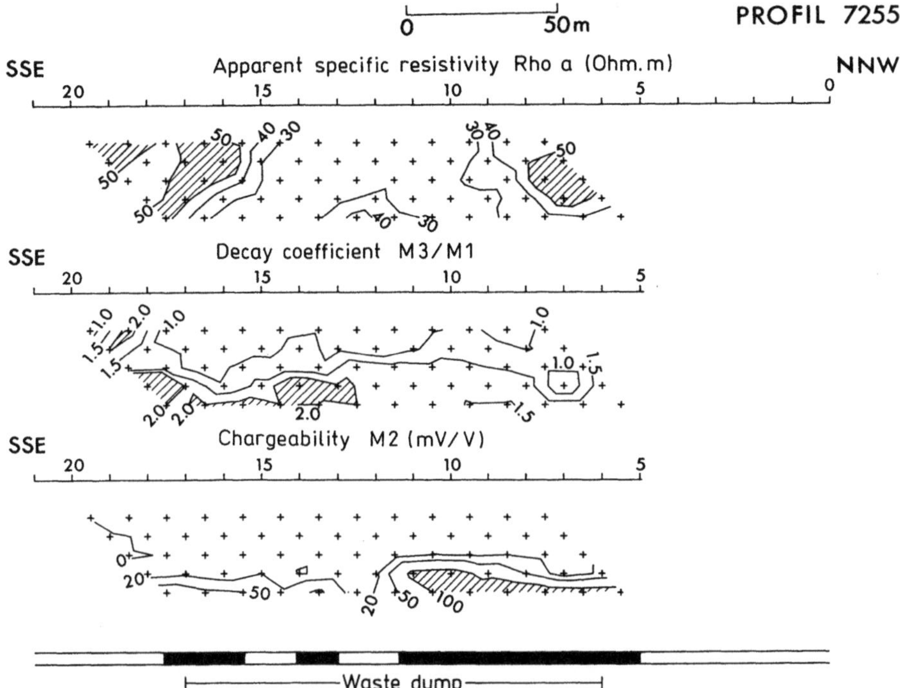

Fig. 3.21. IP pseudosection of hazardous waste site. Dipole-dipole configuration with five different separations (12.5–62.5 m). High chargeabilities >100 mV/V come from galvanic sludge

taic sand, silt and clay, indicated by the resistivities of 75, 20 and 10 Ωm, which are disrupted by faults at the edges of a slump block. The high resistivites >1000 Ωm near the surface are attributed to hydrocarbon residues, visible at the surface and in monitor wells.

This case is an excellent illustration of the great amount of information that can be obtained from simple resistivity measurements by mathematical processing and painstaking consideration of known geological features.

Induced Polarization (IP)

The last two case histories have proven that VES enable the detection and delineation of salt-water intrusions without disturbing sensitive contact zones. However, such conclusions are based only on low resistivities. There are no means of discerning clays of equal low resistivity from saline ground water by DC methods. To overcome this obstacle, the geoelectric method of induced polarization described below should be employed. It determines the chargeability, which is either strong on the surfaces of metals or weak on the boundary layer inside the pore space of rocks. Salinity of fluids within the pore space prevents any charges. Clay, in comparison, possesses moderate chargeabilities.

The technique of this method is more complicated than DC measurements and should mainly be used where electrically chargeable materials or rocks are pre-

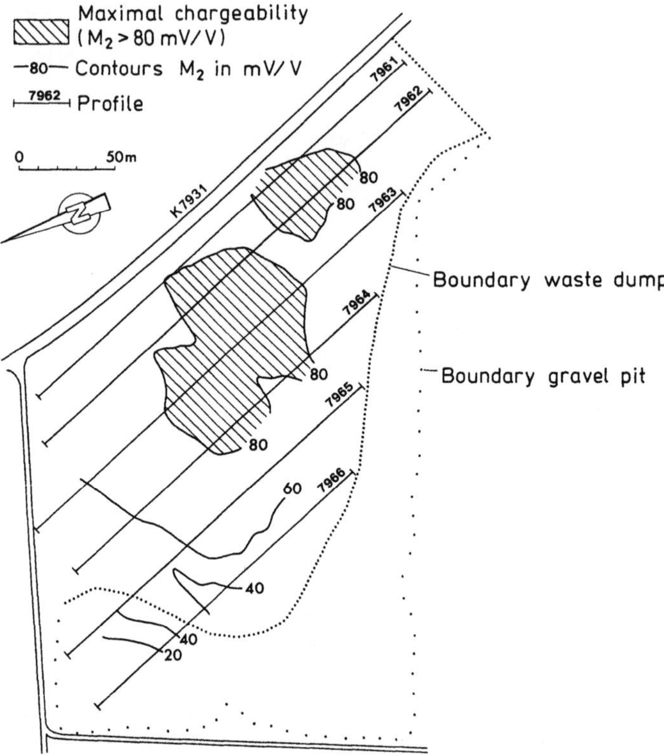

Fig. 3.22. Chargeability contours in mV/V of a domestic waste site in the alpine foreland. Dipole-dipole array, dipole width = 10 m, distance between dipoles = 50 m ($n = 5$)

sent. Known examples are: galvanic sludge, metals, glazed ceramics and printed paper. This list is still very small and research work is necessary to determine the chargeability/resistivity of many more waste components.

The pseudosection of chargeability in Fig. 3.21 shows a completely different pattern from the section of the apparent specific resistivity. The extremely strong chargeability of the galvanic sludges is accompanied only by a weak increase in resistivity from 25 to 50 Ωm. In this and other cases, only the IP method is able to provide knowledge about deposits of special materials within a dump.

The resistivity contours of the domestic waste site in the alpine foreland (Fig. 3.10) evince only one center. The IP map of the same dump (Fig. 3.22) has but two pronounced maxima, marked by chargeabilities >80 mV/V. The IP pseudosection of Fig. 3.23 crosses the southern maximum. The chargeabilities of the dumped domestic garbage are, in contrast to the low resistivities, generally very high, and sink rapidly from 100 mV/V to <20 mV/V upon entering the country rock of gravel and sand. These differences between resistivity and chargeability in location and value prove that the IP anomalies must stem from other sources than the anomalies of resistivity.

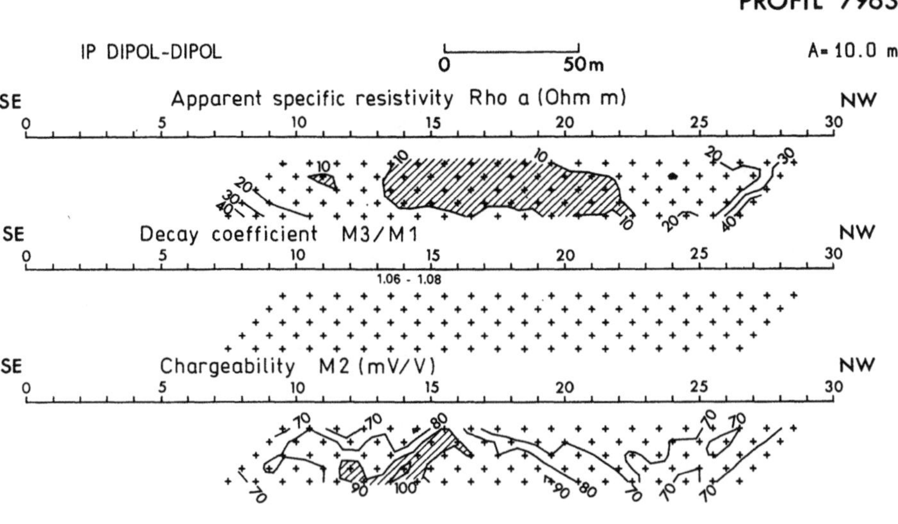

Fig. 3.23. IP pseudosection of domestic waste site Fig. 3.22. High chargeabilities show more details of deposited material than low resistivities

It is not known, though, what kind of deposited material causes strong chargeabilities. To answer this question, it is necessary to determine the chargeabilities and resistivities of different waste types by laboratory measurements and field surveys on deposits of known composition. But it is already possible to discriminate between different waste components of equal resistivity by an IP survey.

Self-Potential

A map of self-potential anomalies (Sp) is presented in Fig. 3.24. They cover the landfill, the magnetic anomalies of which were described in Fig. 3.4. The survey used a grid of 15×15 m². While most magnetic anomalies were found in the northwest part, negative Sp-values down to -80 mV occur in the west of the landfill, positive anomalies up to 55 mV at its western side.

This anomaly pattern neither agrees with the extension of the landfill, nor with the morphology of the site. It is possible that the whole body of the landfill acts as one Sp-source with its plus pole in the west and its minus pole in the east. Nevertheless, no conclusion as to the details of extension of this landfill can be drawn.

This case history portrays the difficult interpretation of environmental Sp-data. The anomaly pattern may also be caused by the buildings in the west and/or by the outflow of leachates in the east.

A contaminated plume within the ground water downwash of a domestic waste deposit was investigated by Sp (Fig. 3.25) and compared to the ammonia content

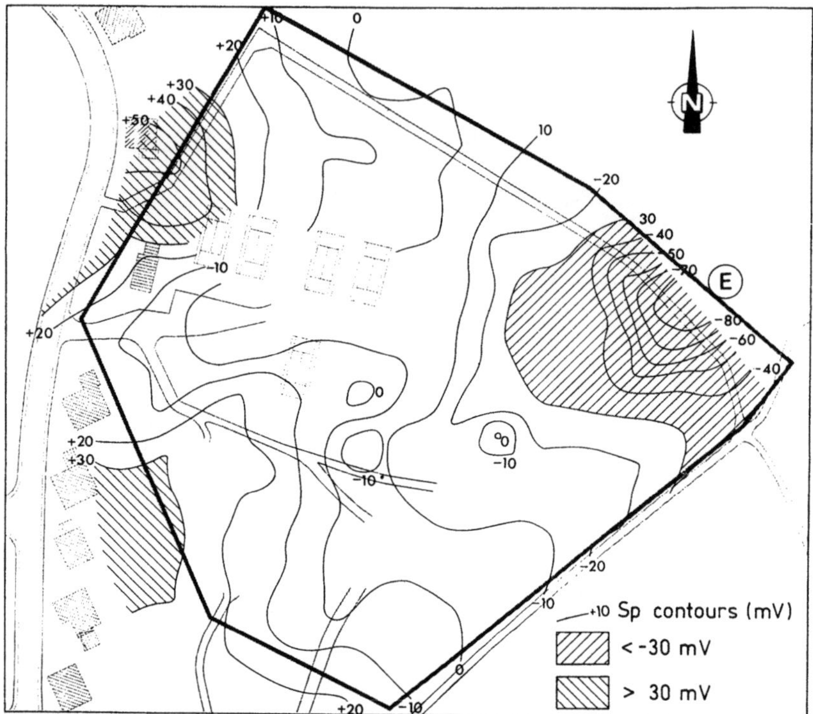

Fig. 3.24. Sp contours of landfill, next to built-up area

of monitoring wells. The Sp measurements followed a grid of $5 \times 2 \, m^2$ and were corrected for spatial noise and time-varying potentials by constant monitoring of a base line.

The Sp anomalies in Fig. 3.25 are regarded as the effects of the near-surface water saturation in marshland. The low Sp values in the middle are caused by a drainage system; the higher potentials left and right occur over water-saturated land. This means that this plume, even with an increased ammonia content, could not be traced by the Sp survey.

Figure 3.26 discloses the self-potential of a Bavarian domestic waste dump. Predominant is a strong but small-scale anomaly on the left side of the dump. It occurs in both cross sections, but its origin is not known. The margin of the waste deposit is visible only in section P3 as a small minimum within a general Sp decline. The cross section P4 is only 25 m away but shows no indication of the lateral extension of the waste.

These obvious limitations of Sp measurements at waste sites are enhanced by the strong influence of frequent local variations of the potential field near industrial areas, which are caused by the switching of high-tension lines or similar man-made processes. Such variations often exceed the measured potentials. They can be eliminated by the continuous registration of a base line, supported by the simultaneous scanning of many potential electrodes.

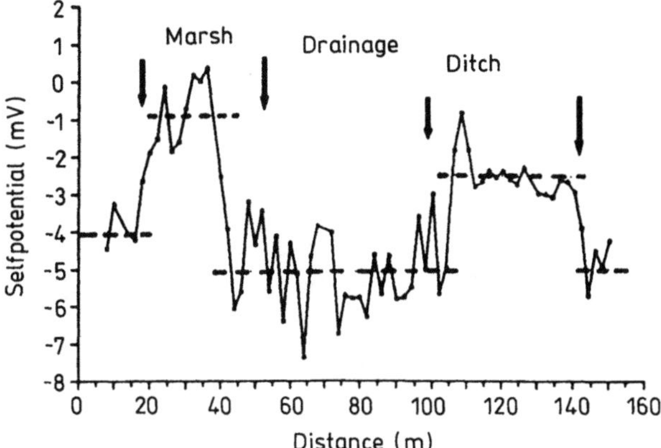

Fig. 3.25. Sp cross section of ground water downwash, domestic dump, Bavaria

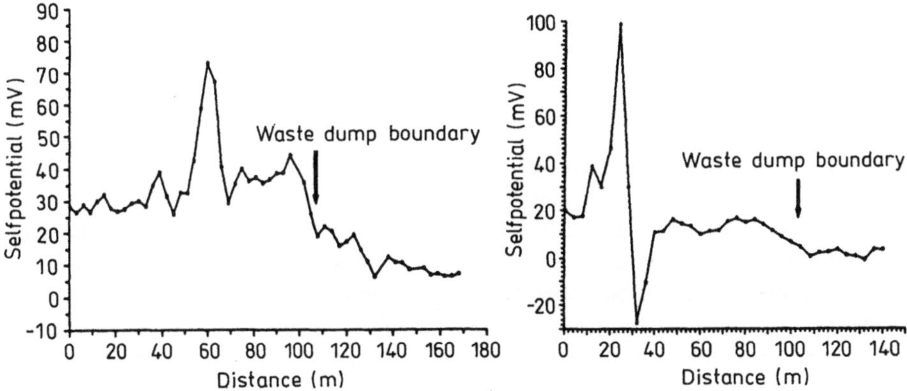

Fig. 3.26. Self-potential profiles across domestic waste dump, feeding ground water to area of Fig. 3.25

3.1.4 Electromagnetic Methods (EM)

Electromagnetic Mapping

EM results are comparable to those of DC mapping. However EM methods progress faster and cost less, because there are no electrodes to be grounded. The evaluation procedure is nevertheless more complicated, since instead of the simple Ohm's law, the complex formulae of Maxwell have to be used. However, there are suitable digital programs for sale, which allow correct and fast interpretation.

In principle, EM mapping records resistivity or conductivity contrasts caused by dumped material or geological structures, similar to the DC mapping. Its

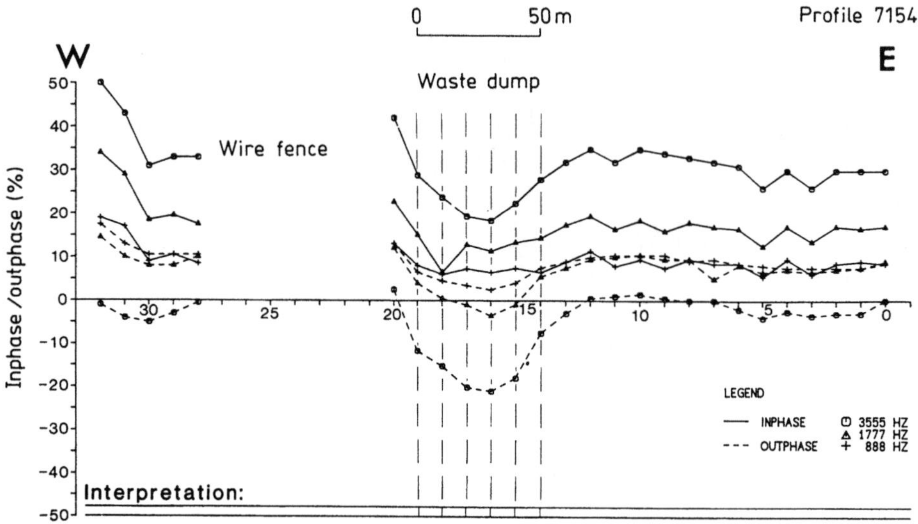

Fig. 3.27. EM cross section: In- and outphases of industrial waste site, contaminated by organics and heavy metals

advantage, however, is better lateral discrimination, especially of steep-dipping but long-stretching small structures. Extensions of hazardous waste dumps, non-iron metals and salty leachates can be easily located.

Its depth penetration depends on the frequency (see Sect. 2.2.2). The frequencies between 800 and 7000 Hz are most suitable. Higher frequencies >15 kHz may not be able to penetrate clayish barriers or seals.

Figure 3.27 presents the result of EM mapping on top of an industrial hazardous waste dump containing heavy metals and CFC. Coplanar transmitting and receiving coils were carried horizontally a distance of 50 m across the dump. The in- and outphase data of the frequencies 888, 1777 and 3555 Hz were registered. The hazardous waste dump stands out clearly as a wide, flat minimum of all phases and frequencies. This is due to the low resistivity of the dumped material, which is still lower than the resistivity of the surrounding clayish sediments.

Though the dump is covered by a plastic liner, there is no visible sign that this insulating foil should influence the EM data: it did not cause the expected increase in EM values. This is due to the measuring distance of 50 m. Either the transmitter or the receiver were positioned outside the small dump at every measurement. If both coils were located over the foil, high resistivities would have been recorded.

The disturbance of EM mapping by artificial metal objects is also illustrated by Fig. 3.27. A wire fence influenced the data so much that a gap of 75 m had to be included.

Figure 3.28 shows another case of EM mapping at a waste site. The dumped material was a mixture of domestic and industrial waste. Coplanar horizontal

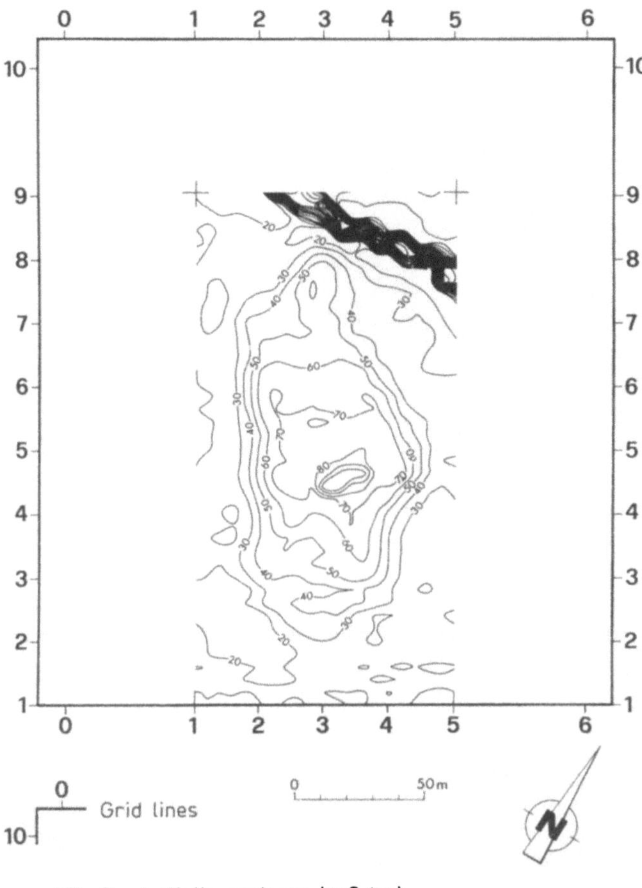

Fig. 3.28. Contours of conductivity (mS/m). Electromagnetic map of a hazardous mixed waste dump

coils were moved at a constant separation of 20 m. The distance of survey points was 5 m. By using the data of six frequencies between 110 and 10 000 Hz, the conductivity in mS/m (millisiemens/meter = the reciprocal of Ωm) was calculated and a contour map constructed.

In these contours, the oval margin of the dump is clearly marked by the 30 mS/m contour. The conductivity increases inside the waste body with >100 mS/m as peak value. Again, the influence of a metallic installation is stronger than the EM values of the dump. This is shown in the northeast corner, where a metal gas pipe is dug into the ground (Fig. 3.21).

The successful application of EM mapping to an abandoned industrial site is illustrated by Figs. 3.29 and 3.30. A 2×2 m^2 grid of 64×60 m^2 was surveyed by the coplanar coil instrument "Maxmin". The coil separation was 10 m; the frequency, 7040 Hz. The contouring of the reinforced concrete foundations of a steel

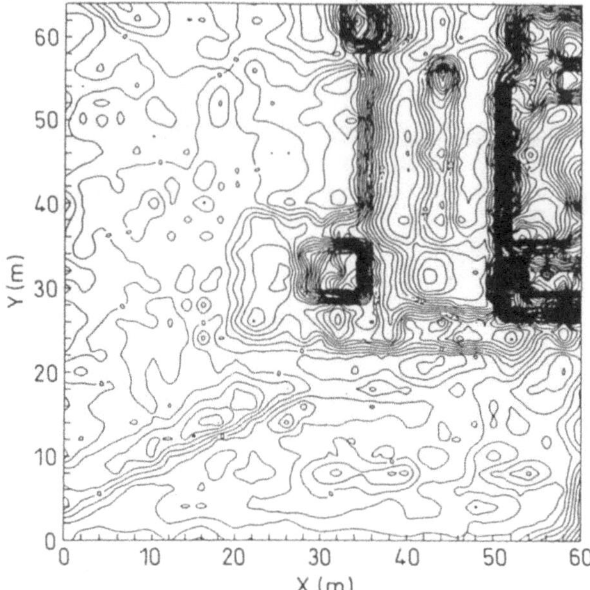

Fig. 3.29. EM contours (inphase, 7040 Hz) of abandoned industrial site. Foundations are exactly presented

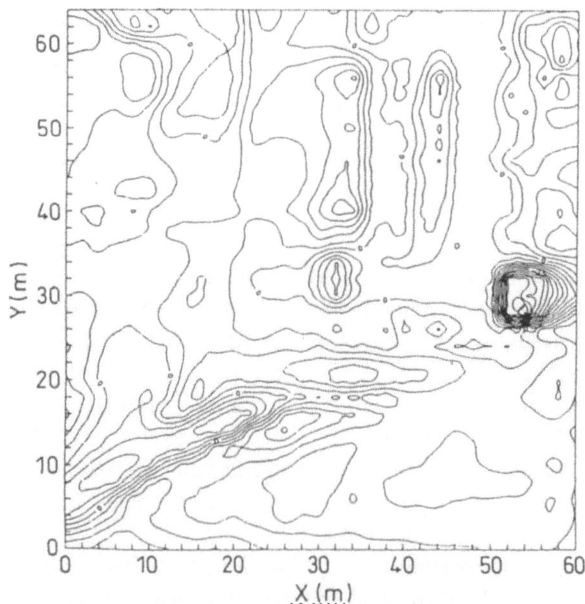

Fig. 3.30. EM contours (outphase, 7040 Hz) of abandoned industrial site, Fig. 3.29. Foundations are less prominent

mill is very precise. Additionally, an underground channel leading to the southwest is clearly visible.

This good result should make it compulsory to survey abandoned industrial sites by geoelectrical, and especially electromagnetic, methods. This will improve the detailed planning of remedial work and save considerable costs.

In Figs. 3.31 and 3.32, an EM inphase map with a grid of 2×2 m^2 is laid over a subterranean air-raid shelter, using the frequency of 3500 Hz. It was suspected that several big metal containers were hidden inside the shelter, but the magnetic survey (Fig. 3.6) revealed only the outline of the shelter. However, the EM survey penetrated the thick roof of reinforced concrete and resulted in a trifold structure. This is interpreted with great probability, as an EM response to the containers with contaminated fills.

This case illustrates once more the better resolution of EM mapping compared with a mere magnetic survey. Nevertheless, only the combination of magnetic and geoelectric or other methods will provide all the details necessary to properly mitigate or remedy contaminations.

The EM mapping (frequency domain) of a plume of a hazardous waste site in the USA is demonstrated by Figs. 3.33 and 3.34. The continuously recorded EM data were obtained at 6-m and 16-m depths, using two separate EM systems. Thirty parallel profiles, each 1000 m long and spaced by 30 m, were measured.

In spite of the natural variation of the conductivity (inversed resistivity) by the geology of the area, the conductivity contours and the three-dimensional picture clearly portray the lateral extension of two contaminated plumes, to the west and to the northeast, plus the location of monitor wells by minor lobes of anomalies.

A saline ground water plume was mapped by the electromagnetic instruments Geonics EM 31 and EM 34 in Merced County, California USA (Figs. 3.35 and 3.36). Its chemical contamination comes from several ponds of a 1280-acre reservoir into which agricultural drainwater containing traces of selenium and other toxic elements had been discharged. The purpose was to outline the underground distribution of selenium contamination, which had already caused deformities of birds (waterfowl) and might eventually enter the human food chain.

To remedy the area, the ground water contamination had to be precisely located. It was necessary to differentiate between surface inhomogenities and the saline plume at a depth of 5 m to 20 m below surface. To achieve this, three coil separations of 3.7 m, 20 m and 40 m were used in a survey covering 2 km^2. Though the native ground water is already saline and has conductivities ranging from 300 mS/m to 400 mS/m, the conductivity of the contaminated ground water falls to <1000 mS/m, due to the increase of Na, Cl and SO$_4$.

The smallest coil separation 3.7 m (Fig. 3.35a) of EM 31 responded most to topographic features and variations in soil moisture and salinity, caused, for instance, by dry lake and creek beds. By increasing the intercoil separation of EM from 34 to 40 m, those surface effects diminished (Figs. 3.35b and 3.36a). Finally, an interpretation map was constructed from the EM data (Fig. 3.36b).

The leading edge of the plume has migrated up to 350 m from the discharge ponds (areas B and D in Fig. 3.36b). At pond 1, where the release of saline drain-

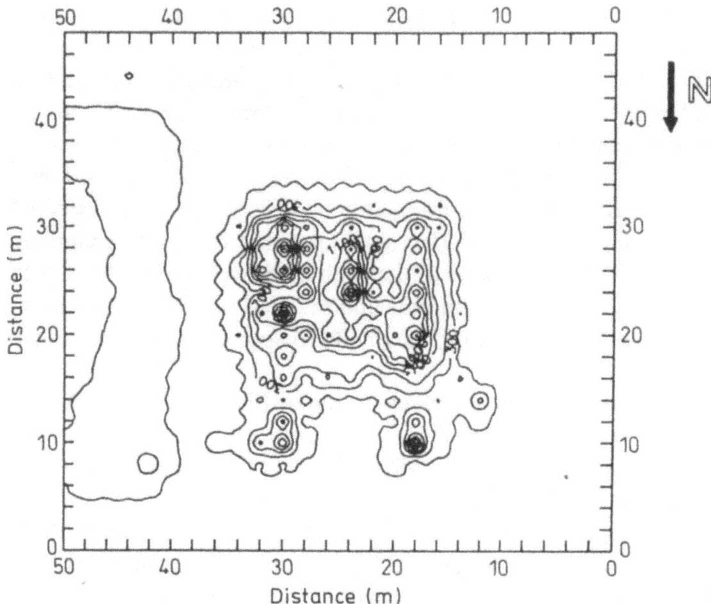

Fig. 3.31. EM contours (inphase, 3500 Hz) of buried air-raid shelter with reinforced concrete roof and walls (Fig. 3.6)

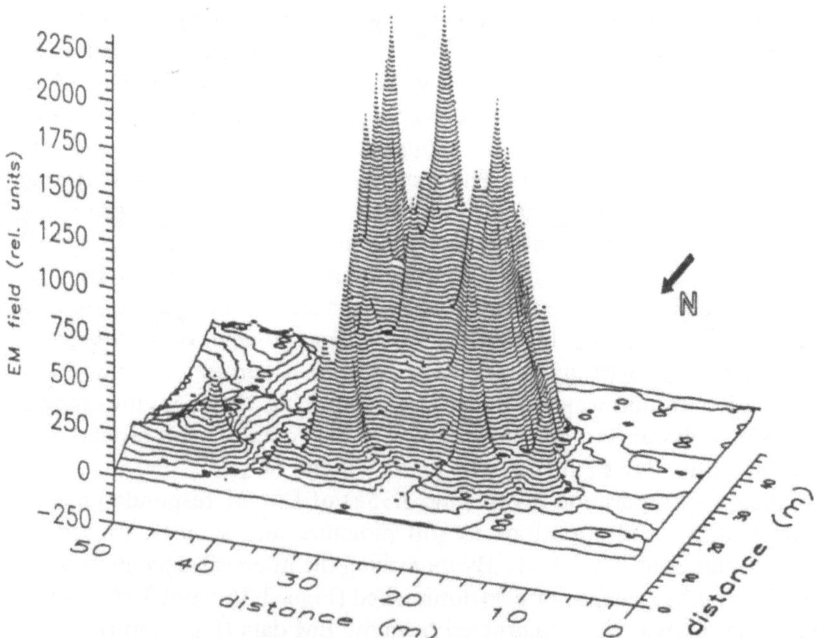

Fig. 3.32. 3D presentation of Fig. 3.31. Three peaks disclose three metal containers, filled with toxic liquids

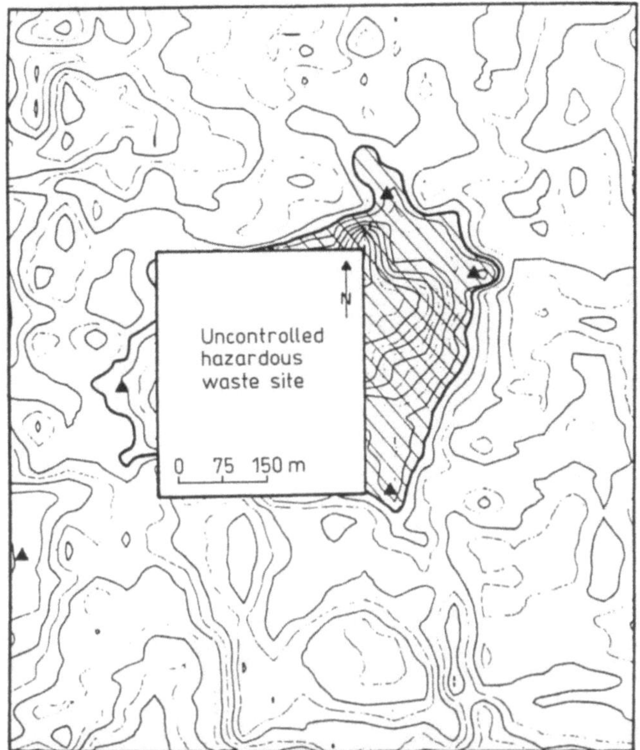

Fig. 3.33. Contours of EM conductivity data at a hazardous waste site with two plumes and monitor wells (triangles)

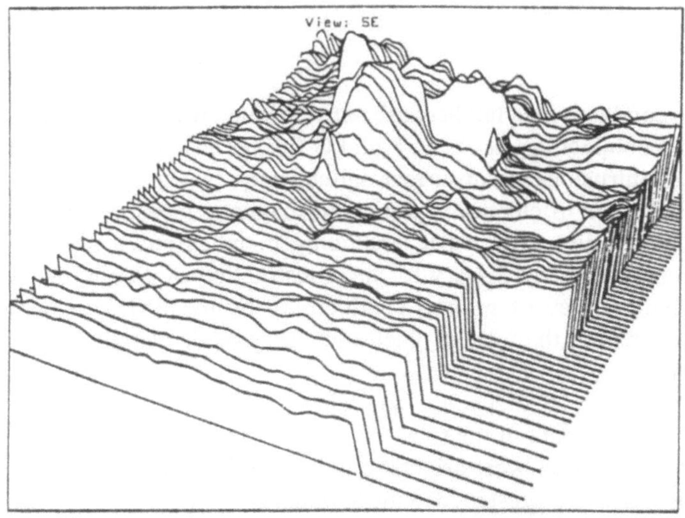

Fig. 3.34. 3D picture of EM data, Fig. 3.33. Low-conductivity plumes differ clearly from geological noise

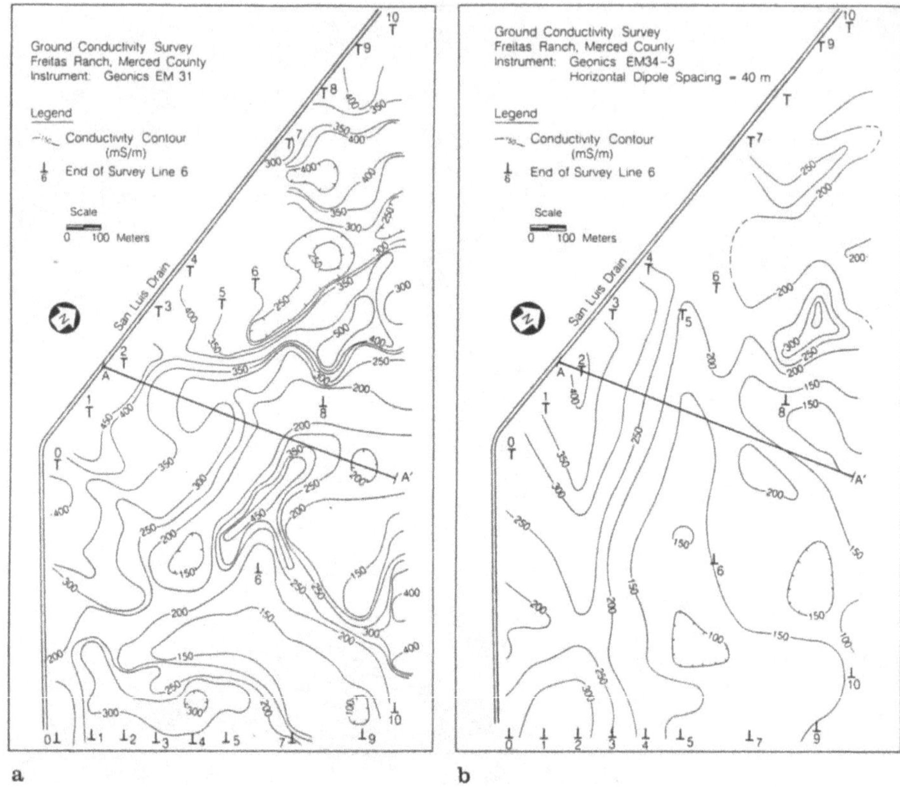

Fig. 3.35. Contours of equal conductivity of plume adjacent to the Kesterson Reservoir, Merced County, California.

water had ceased, the conductivities fell because native ground water had displaced the plume. The low conductivity areas C, F and G further east are interpreted as soil salinization and intrusions of subsurface saline water.

The combined application of electromagnetic and magnetic measurements is described by Fig. 3.37. Buried paint waste, including metal drums, had to be located rapidly. Its removal was necessary for the construction of a building. A shallow electromagnetic survey of ground conductivity by the EM 31 m was amended by measurements of the total magnetic intensity made using a proton precession magnetometer.

A rectangular grid of 20×20-ft spacing was laid over the suspicous area of 400×700 ft. At 762 stations, data of the total magnetic intensity and of the conductivity in north-south and east-west orientations were collected. In Fig. 3.37, the resulting contour maps are presented with anomalies identified by capital letters. The raw magnetic data of Fig. 3.37a have been corrected for the magnetic field, created by a building in the east. The contours of conductivity were con-

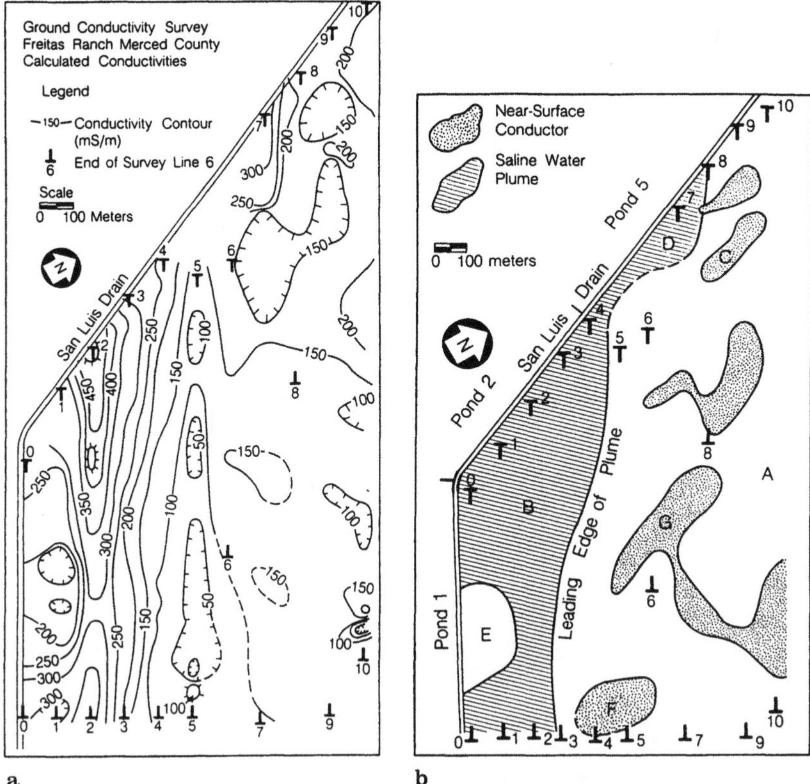

Fig. 3.36. Evaluation of EM measurements at Kesterson Reservoir (Fig. 3.35).

structed from the N-S-oriented EM measurements in Fig. 3.37b. This approach allowed the classification of the waste into 11 categories in Fig. 3.38.

Ponds are filled with saline drainwater. In Fig. 3.35a, a 3.7-m coil separation reveals surface salinity. In Fig. 3.35b, a 40-m coil separation reports salinity of plume penetrating 20 m.

As regards profiles of ground conductivity, Fig. 3.36a discloses a rapid decrease in conductivity away from the reservoir and a new increase ~700 m east. Figure 3.36b shows several origins of increased conductivity/salinity. Areas B and D house plumes; C, F and G contain saline surface water

The successful verification of the geophysical results by follow-up drilling, excavation and trenching is tabled in Fig. 3.38. This detailed list of the sources of anomalies is based on the excavation work and proves the great advantages of geophysical work at hazardous waste sites. Without geophysics, the whole building prospect would have been delayed because of the necessary drilling, probing and trenching activities. This would also have resulted in a great increase in cost.

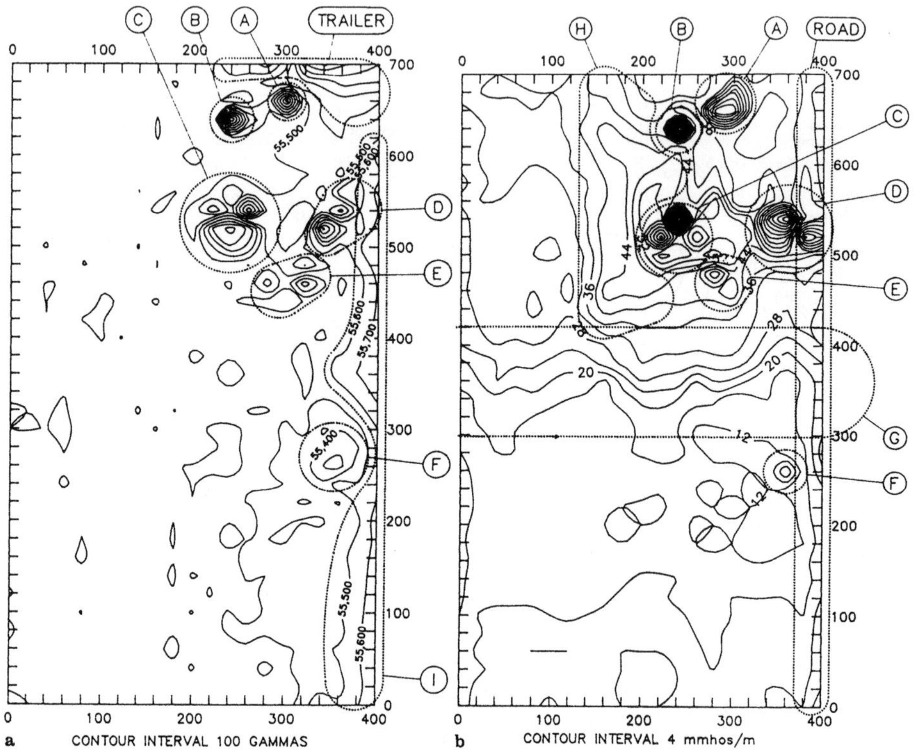

Fig. 3.37a, b. Magnetic and EM contours of scattered hazardous waste. **a** Total magnetic field. **b** Conductivity by EM survey (instrument axis N-S)

Transient ElectroMagnetic Soundings (TEM, TDEM)

The response of a contaminated, slightly saline plume to transient electromagnetic methods has been tested at a domestic waste disposal site near Perth in Australia. This site was very suitable, since the general spread of this plume was already known by a dense drilling program.

In Fig. 3.40a, the dipole receiver was placed in the center of a transmitter loop 25 m in diameter. In order to define the complete extent of the leachate from a pit of solid waste, 83 TEM soundings were made over a period of 6 months. The soundings were extended beyond the known extension of contamination to monitor the resistivity background.

TEM sounding is cumbersome because a large loop has to be laid on the ground at every point of observation, but its result is most rewarding. The resistivity contours in Figs. 3.39 and 3.41, which were derived from TEM data, agree with the evidence of pollution by drilling of the holes M4, M5, M6, M7, M9 and M11. Since the drop in resistivity is less prominent than at other contaminated

Anomaly Location[1]	Maximum Instrument Response Above Background		EM Difference N-S less E-W Orientation	Does Response Meet Statistical Criteria?		EM Difference ≥ 3.3 mmhos/m	Physical Description Based on Excavation Work[3]
	EM (milliMhos/m)	Mag (gamma)		EM (10 mmhos/m in north, 4 mmhos/m in south)[2]	MAG (150 gammas)		
A	28	250	28	Yes	Yes	Yes	Waste pit: 12' x 50', 6' depth, 2' below surface containing two drums.
B	40	900	4	Yes	Yes	Yes	Waste pit: 15' x 25', 7' depth 2' below surface, no drums.
C	23	500	12	Yes	Yes	Yes	Waste pit: 50' x 60', 3' - 5' depth, 1' below surface, some drum pieces.
D	37	700	14	Yes	Yes	Yes	Two waste pits: 20' x 60' and 30' x 40', 3' - 5' depth, more scrap metal, eastern pit: 25 drums, scrap copper wire and 5-gallon cans.
E	11	400	8	Yes	Yes	Yes	Waste pit: numerous pits 5' - 10' diameter, 4' depth, 1' below surface, some drums, numerous 5-gallon cans.
F	13	200	8	Yes	Yes	Yes	Tin battery remains: 7' x 15', 4' depth, 2' below surface, sheets of tin, lumber, nails, concrete; no waste.
G	15	0	0	Yes	No	No	Geologic feature: bedrock; slopes to the South.
H	12	0	0	Yes	No	No	Tree Root Zone: Aerial photo shows wooded zone between disposal area and farm field. 4' depth, 2' below surface.
I	0	900	0	No	Yes	No	Man-made feature: building 50' east of survey area.
J	0	0	4	No	No	Yes	Concrete slab: 3' x 5' x 6", 2' below surface, contained 1/2" copper pipe; used as a ground in prior field office.
K	0	0	4	No	No	Yes	Surface material: no subsurface feature found.

[1] Location as shown on Figures 3,37

[2] Statistical Criterion: > 1 standard deviation response.

[3] Waste pits contained an assortment of waste, including ash, paint sludge, miscellaneous scrap metal, and occasionally partially intact 55-gallon drums.

Fig. 3.38. List of anomaly sources of Fig. 3.37

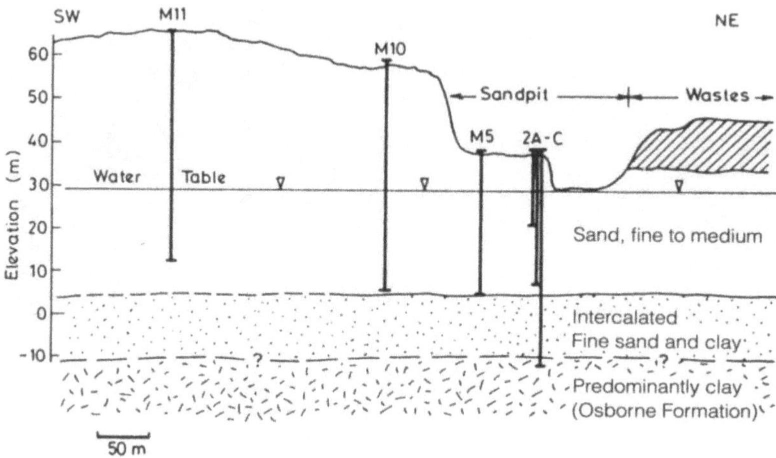

Fig. 3.39. Cross section of domestic disposal site, discharging seepage into an aquifer near Perth, Australia. Ground water flow NE to SW

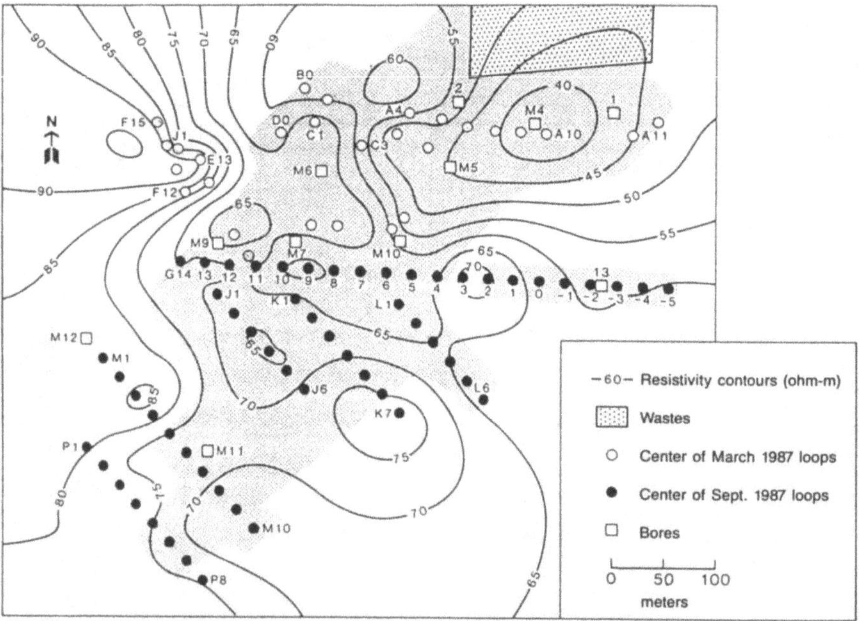

Fig. 3.40. Resistivity contours by TEM soundings of plume in sandy aquifer, next to landfill (Fig. 3.39). Values < 75 Ωm (stippled area) mark spread of the plume

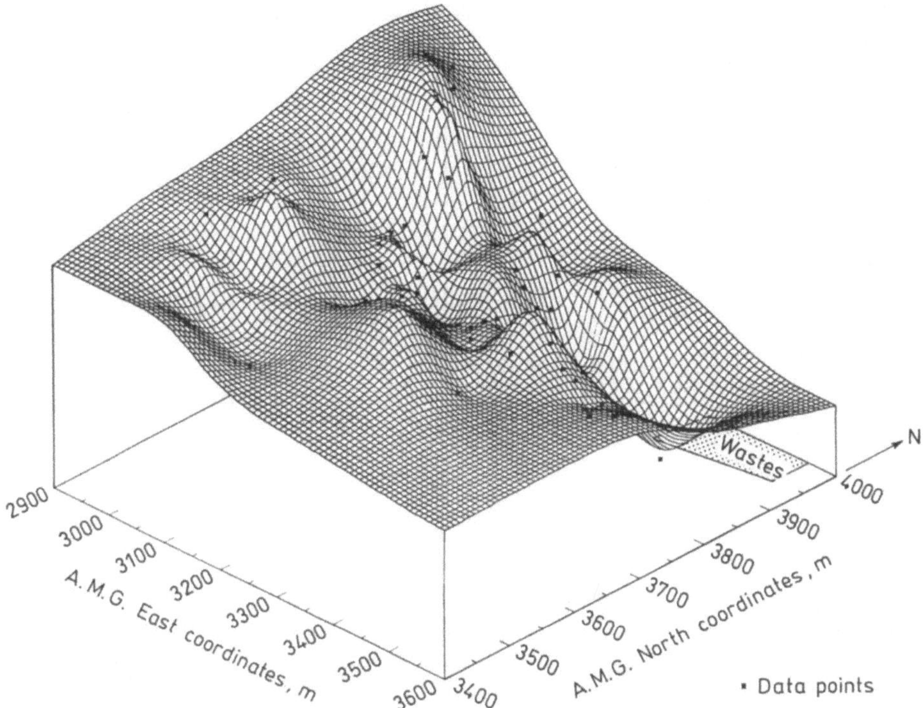

Fig. 3.41. 3D plot of aquifer resistivity near waste site of Figs. 3.39 and 3.40

plumes, this detailed evaluation must be regarded as another successful application of geophysics to a complicated environmental problem.

A more intense brine pollution was explored by TDEM electromagnetic soundings near Pawhuska, Oklahoma USA. The brine was pressed into the ground by injection wells. Numerous metallic pipelines in the field necessitated the wire loops being laid out in an irregular pattern. Two injection wells were located at the TDEM stations 3 and 1 (Fig. 3.42). The second well also lies close to station 18.

The transient soundings were evaluated by inversion and a layered-earth model was constructed. Figure 3.42 shows that the horizontal stratification was interrupted by the steep-dipping margins of brine pollution with the very low resistivities of 0.3 to $4\,\Omega$m. Nevertheless, even the somewhat erroneous picture of Fig. 3.42 conveys important information as to the spread of brine at a depth of >100 m.

Georadar

This relatively new method is impressive by virtue of its technical and electronic perfection and of the presentation of its results in colored sections, already available in the field. Its main difficulty is the strong influence of ground moisture

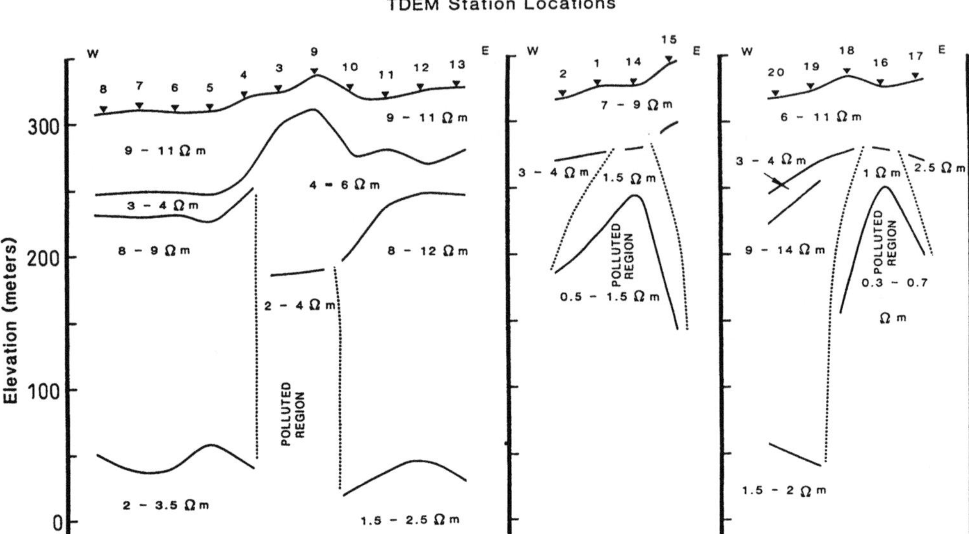

Fig. 3.42. TDEM survey of brine pollution by injection wells, Oklahoma, USA. Injection wells are between stations 3–9, 1–14 and 18–16

on the results. Ground radar sections, obtained before and after heavy rainfall, will differ considerably in depth penetration and pattern of reflections.

Another problem is created by lateral changes in the clay content of overburden and soil. Higher clay content will abruptly minimize the depth of measurement and may give the impression of an upturn in the reflecting horizons. Therefore, interpretation of radar data must be made with great care because the danger of overinterpretation is immanent with this method.

Ground radar was employed to investigate abandoned gasworks (Fig. 3.43). The built-up area was investigated by radar profiling with a line spacing of 0.5 m. The transmitted frequency was 300 MHz. In spite of the small free space between the industrial buildings, this method had good results. A number of single objects could be located that had not been found by geomagnetic measurements (Fig. 3.3).

A special advantage was the detection of linear structures, related to ceramic pipes and drains by radar, which could neither be found by geomagnetics nor by low-frequency electromagnetics.

Since the dielectric constants of organic chemicals are very low (see Table 2.3), ground radar is well cut out to search for organic contaminants, especially tar, volatile hydrocarbons and used oil. Unfortunately, very similar reflections may be obtained from sand lenses or other stuctures of high resistivity. Therefore, partly known organic contaminations may be followed up to establish, for instance, their spatial extension. However, assuming the presence of hydrocarbons, etc., from radar reflections only should be avoided.

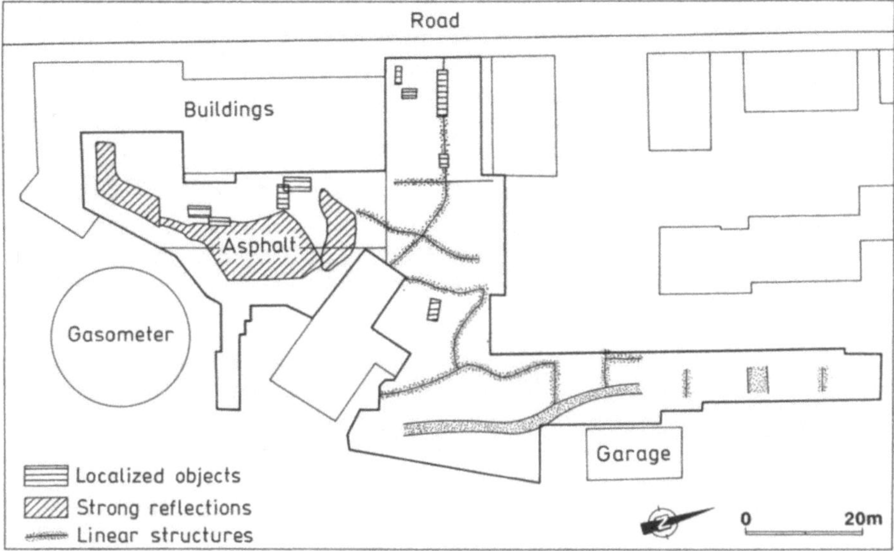

Fig. 3.43. Ground radar investigation of abandoned gasworks

The radar reflections of Fig. 3.44b are clearly effected by the immersion of 770 L of volatile hydrocarbons, shown in the lower radargram. But a congruous pattern could be produced by a pocket of high-resistive material, like dry sand.

The radar record of a petroleum pipeline spill near Bemidji, Minnesota is illustrated by Fig 3.45. The normal texture of the soil is replaced by wholly different reflections on the right, owing to a strong change in resistivity and dielectric permittivity, caused by oil floating on the water table. However, such a featureless texture could also have been produced by dry sand or other stratified material of high resistivity.

The darker left side of the 80-MHz radargram exhibits larger contrasts due to the varying water content of different horizons of a glacial outwash. The depth scale on the left is approximate and not reliable for follow-up activities.

The detection of cavities, especially in abandoned industrial areas, is an important geophysical task that is best fulfilled by ground radar profiling. Yet this is successful only at near-surface cavities, and even this fails if the overburden is clayish and difficult to penetrate by radar signals.

Cavities stand out as areas of increased signal enhancement and depth penetration. In Fig. 3.46a, tunnel is indicated by the sudden augmentation of the number of reflections and the prolongation of the travel time from 50 to 150 ns (nanoseconds).

Ground radar may be used to solve many other environmental questions, provided a feasible relation of the assumed object depth to the depth penetration of radar waves is maintained. It is sometimes possible to attribute singular reflec-

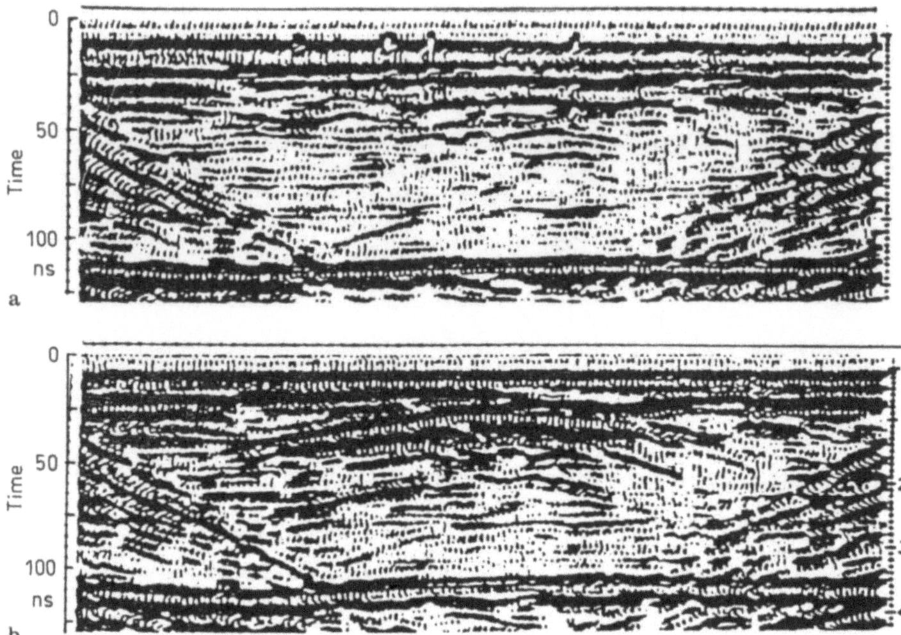

Fig. 3.44a, b. Alteration of radargrams by immersion. **a** undisturbed condition. **b** radargram 16 hours after immersion of 770 L volatile hydrocarbons into sand (frequency 200 MHz.)

tions to a certain structure, if complicated procedures of evaluation have been applied. Such evaluation programs, like migration, may be directly taken off seismic software.

Airborne Electromagnetics

Electromagnetic air surveys are mostly combined with magnetic and radiometric measurements. The magnetic helicopter survey, described in Fig. 3.9, was augmented by the DIGHEM observations of Fig. 3.47. The helicopter flew at an average height of 50 m above the ground. The piled-up dump of domestic waste caused a reduction in the specific resistivity from $>10\,\Omega$m to $<2\,\Omega$m, as the values decreased in concentric lines towards the minimum of the center. The $10\,\Omega$m contour tallies with the rim of the pile.

Compared with the airborne magnetic contours of Fig. 3.9, the EM picture is easier to interpret and the extension of the waste dump is more visible. Needless to say, not only piled-up but also buried waste deposits can be localized and delineated by EM helicopter surveying.

EM helicopter surveys were also successful at mapping saline ground water. The river Weser carries a heavy freight of brines through the city of Bremen, Germany. The brines are discharged 500 km upriver by potassium mines and lower the specific resistivity of the river water to $<6\,\Omega$m. The EM air survey has traced

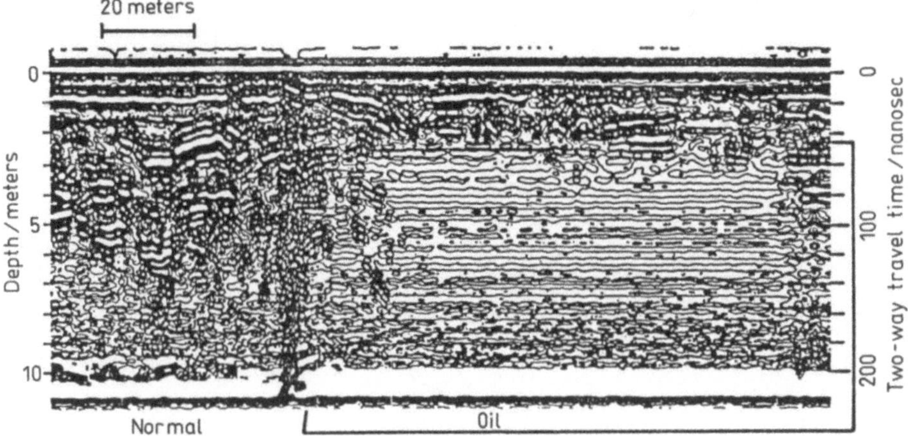

Fig. 3.45. Radargram of a petroleum pipeline spill

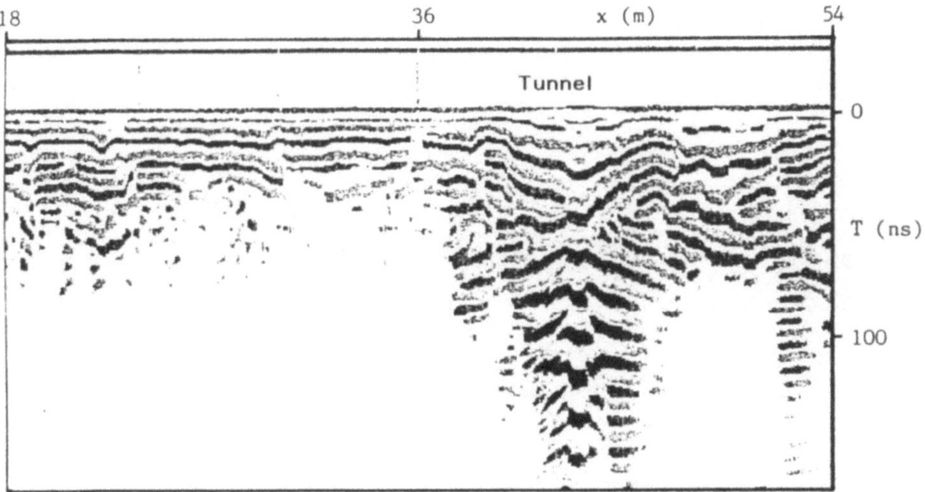

Fig. 3.46. Ground radar cross section of a shallow tunnel (frequency 80 MHz)

several salty plumes with resistivities below 6 Ωm (Fig. 3.48), which spread sideways from the river bed into the upper aquifer. This sand/gravel bed provides fresh water for the city.

Evaluation was checked by mapping the chloride content of the upper aquifer by flush drilling. The drilling of many holes took 2 years; the helicopter flights, however, were accomplished within 2 days. Astonishingly, both methods had the same results!

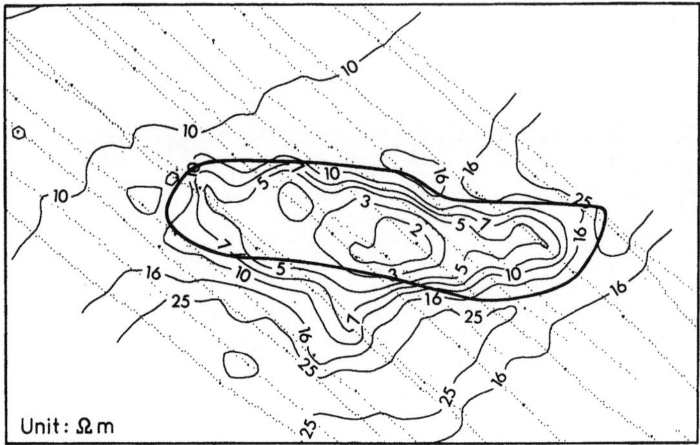

Fig. 3.47. Contours of approx. specific resistivity by EM (DIGHEM) helicopter survey of the domestic waste dump of Hannover, Germany (Figs. 3.9 and 3.61). Flight line spacing ~50 m, frequency 385 Hz

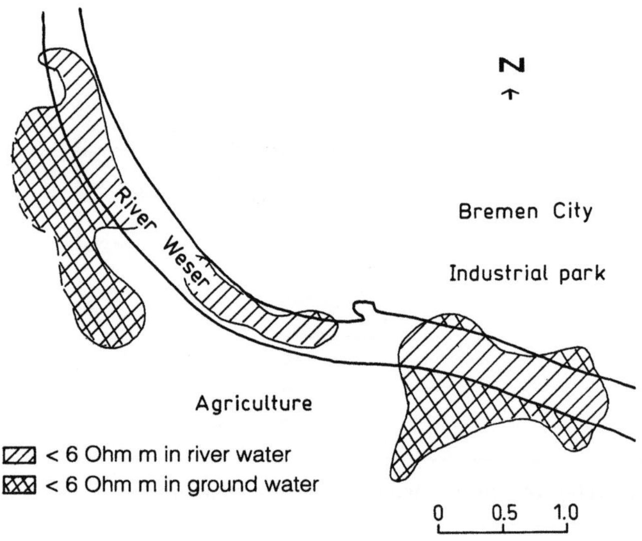

Fig. 3.48. EM helicopter survey of saline seepages from the river Weser, Bremen, Germany

3.1.5 Seismics

Seismic Refraction

The refraction method is able to detect flat-lying boundaries between beds showing different seismic velocities. A precondition is an increase in seismic velocitiy in the lower horizon. It functions well at shallow depths and is often employed to

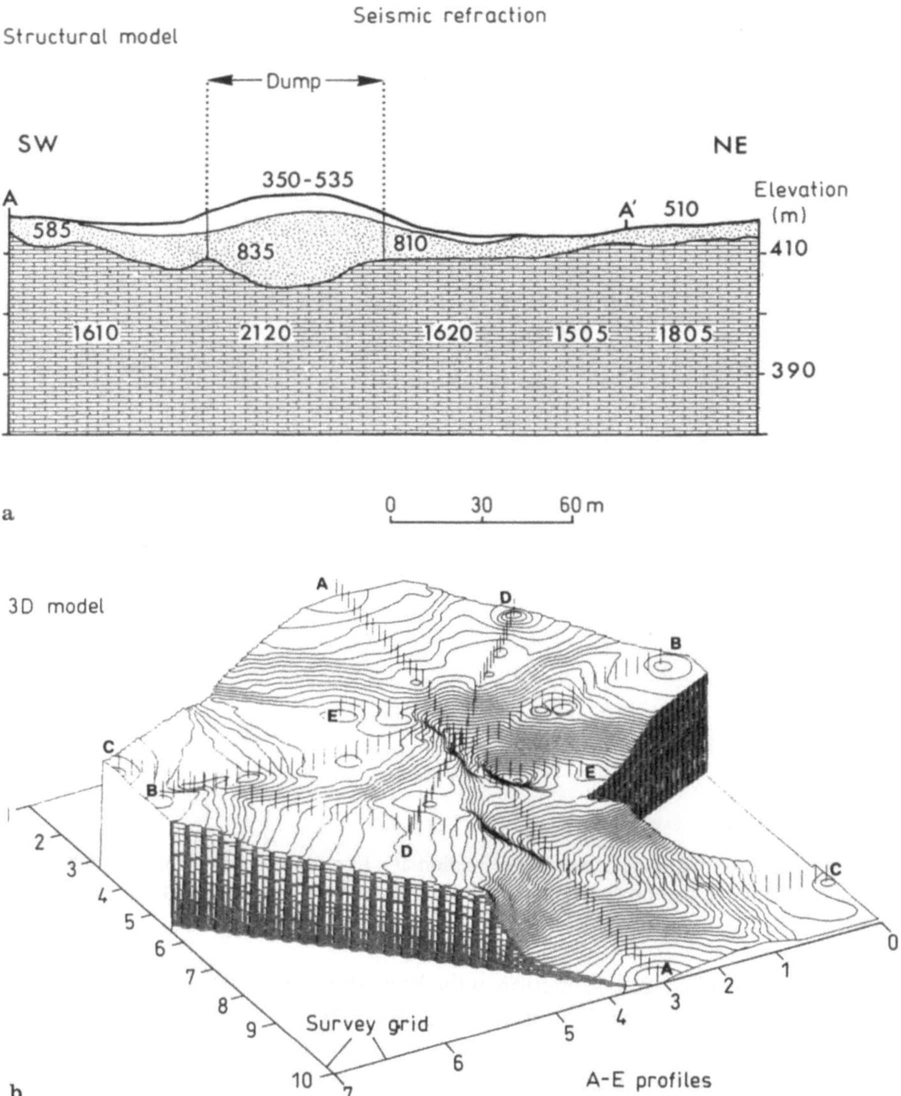

Fig. 3.49a, b. Results of seismic refraction at a hazardous waste site. **a** Structural model with seismic P-velocities [m/s] along cross section B; **b** Three-dimensional picture of the refractor surface

work out the thickness of unconsolidated layers or the relief of the basement rock. Underground depressions, cut into the surface of impermeable hard rock, may guide contaminated seepages or leachates and can be traced by seismic refraction.

The borders between hazardous waste dumps and unconsolidated sediments are rarely found by seismic refraction because the seismic velocities of waste and unconsolidated rock are too similar.

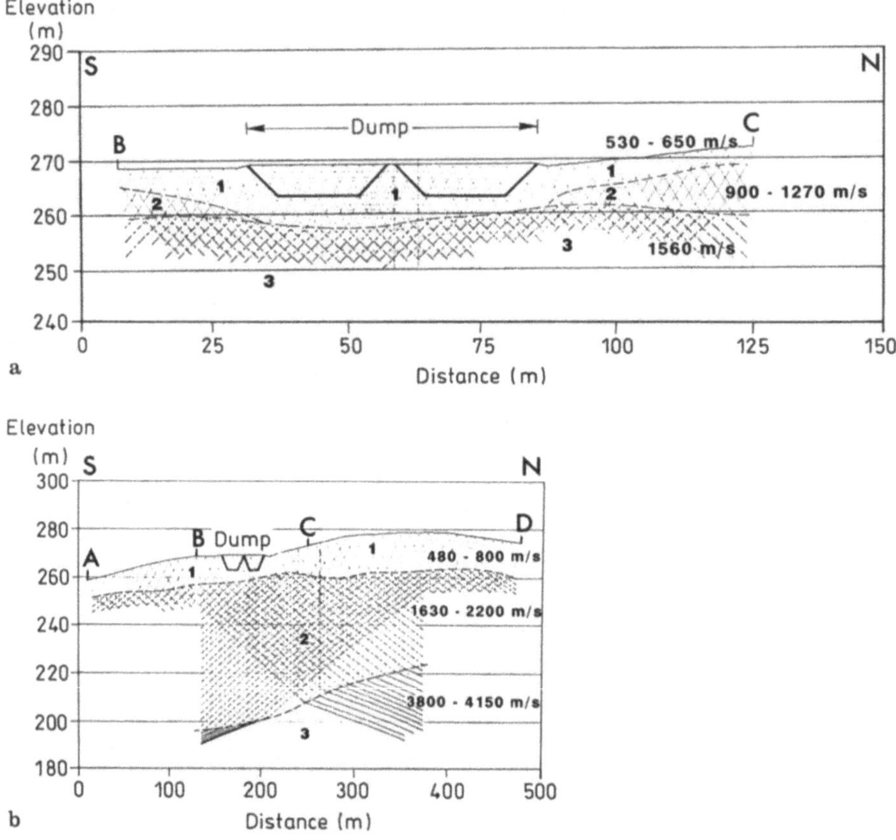

Fig. 3.50a, b. Cross sections of seismic refraction at a hazardous waste dump with hydrocarbon contamination. **a** section geophone, spacing 2.5 m; **b** section geophone, spacing 10.0 m

Five sections of seismic refraction at the hazardous waste site of Fig. 3.11, containing a mixture of hazardous industrial and domestic disposals, revealed detailed information about the relief of the surface of the sandstone refractor. Figure 3.49a depicts a typical section. The surface of the sandstone is presented in the three-dimensional picture of Fig. 3.49b. It gives clear indications of the depressions within the sandstone, which are the outlets of contaminated seepages.

In this case, the refraction method met with favorable geological conditions: the seismic velocities of the unconsolidated and weathered upper bed vary from 510 m/s to 835 m/s. The velocities of the sandstone with reduced hydraulic conductivity are much higher; they range from 1500 m/s to 2100 m/s.

However, the velocity contrast (810 m/s to 835 m/s) between dumped waste and the weathered upper bed is too small to map the extension of this hazardous waste site by seismic refraction.

The next case (Fig. 3.50) concerns an industrial hazardous waste dump whose electromagnetic survey was described in Fig. 3.27. The seismic refraction used

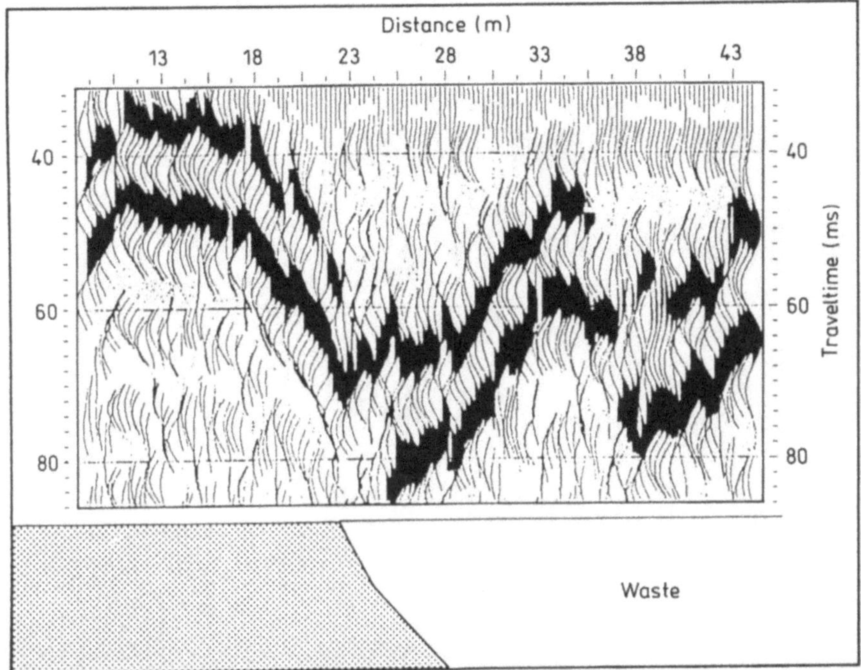

Fig. 3.51. Locating the border of buried waste by air-acoustic seismics (Tekoni Innovations)

two different geophone spacings of 2.5 m and 10 m. Again, the border of the dump and its base did not show up in the results of refraction; the contrast in the velocity of seismic waves was too small.

Three layers were established:

Layer 1: weathered topsoil 530 m/s – 800 m/s,
Layer 2: weathered hard rock (clay with gypsum) 900 m/s – 2200 m/s,
Layer 3: hard rock (clay with gypsum) 1560 m/s – 4150 m/s.

The seismic velocities are surprisingly much higher in section b with 10 m geophone separation than in section a with 2.5 m (Figs. 3.50a and 3.50b). Furthermore, the lowest reflector, expected at a depth between 50 m and 70 m, seems to be inclined in the longer geophone distance. It is interpreted as the table of gypsum at the border between leached and compact rock.

This case shows that non-horizontal bedding, caused by leaching or subrosion of gypsum from below, may jeopardize evaluation if the geophone separation is too small.

Air-Acoustic Seismics

Figure 3.51 shows the result of air-acoustic mapping across the border of a waste dump. The distance between the loudspeaker as seismic source and the geophone

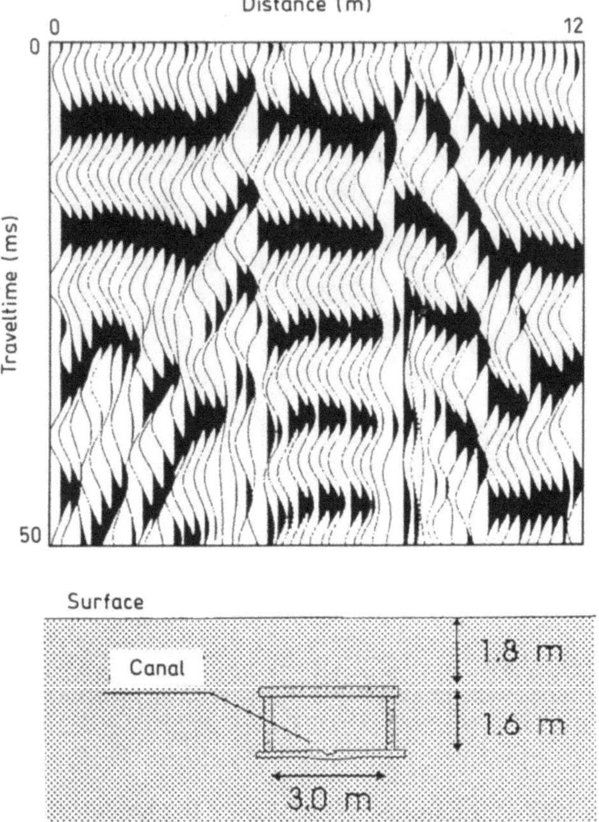

Fig. 3.52. Detection of a buried sewage canal by air-acoustic seismics. Distance loudspeaker/ source to geophone 1 m, separation of stations 25 m, frequency of seismic signal 200 Hz (Tekoni Innovations).

was 8 m. The seismic stations were separated by 25 m, and the average signal frequency was 300 Hz.

While short and almost equal travel times dominate the left side of the section, originating from clayish rocks, the travel times increase toward the right. The margin of the waste deposit is clearly visible by this change.

Air-acoustic seismics are one more tool to locate the extension of buried waste, to be applied especially if magnetic and geoelectrical methods fail.

Even the difficult search for underground cavities can be performed by air-acoustic seismics. Figure 3.51 proves that the walls of an underground canal can be located by this method. They are clearly marked by an anomalous decrease of the travel time. Since the same effect was observed on parallel sections, the length of the drain could be found from the surface. The method was able to find similar structures down to 13 m of overburden.

Seismic Reflection

Exploration of oil and gas is not possible without seismic reflection. Therefore, this method is used in more than 90% of global geophysical activities. In solving environmental problems, seismic reflection is, however, less prominent. One reason is its penchant for great depth penetration and neglect of near-surface effects. Another cause is the high cost of reflection surveys.

New digital instruments with high-frequency sources and extremely fast sampling rates for geophones have been developed, and now allow evaluation of reflections from depths < 50 m.

Nevertheless, even with this new development, seismic reflection is best suited for structures that lie below the level of most hazardous waste sites. Such tasks as establishing the location, thickness and extension of aquifers and tectonic faults, which guide the underground movement of dangerous liquids, can be well tackled.

The seismic record of Fig. 3.53 was obtained from below the hazardous waste dump described in Fig. 3.50. In contrast to the results of refraction, there is no indication of the 20-m thick layer of low velocity at the surface. In spite of its narrow geophone spacing of 10 m, the seismic evaluation in Fig. 3.53a also contains no trace of the border of the buried waste site.

In the seismogram of Fig. 3.53b, a strong reflector starts in the north at a depth of 45 m and sinks down to 70 m. From there to the south, the reflector seems to be broken up into three short horizontal pieces. This structure is interpreted as a series of faults having downthrusted the boundary between clay and banked limestone.

A borehole was sunk into this fault zone and a pumping test found high permeability in fractured limestone. In spite of the fact that the seismic reflection did not convey any features of the hazardous site, it was worthwhile employing it because it depicted a hydraulically active fracture zone at greater depth. This zone may allow the movement of contaminated ground water into a lower hydraulic level, which feeds the waterworks of a neighboring town.

This case history conveys that seismic reflection, despite its limitations at shallow depths, can be a sensible tool to solve hydraulic environmental problems. The information from depths > 50 m regarding horizontal stratification and its interruption by faults cannot be obtained so clearly by any other geophysical method.

Figure 3.54 reports a combination of seismic refraction and reflection on a 700-m profile at an industrial hazardous waste site. The seismic model distinguishes four beds:

1. unconsolidated soil and loam   480 – 750 m/s,
2. unconsolidated silt and clay   1550 – 1800 m/s,
3. banked limestone   2200 – 2500 m/s,
4. massive limestone   3900 – 4200 m/s.

The position and the thickness of the waste deposit in Fig. 3.54 is again not discernible from the seismic data. The ground water table was registered in bore-

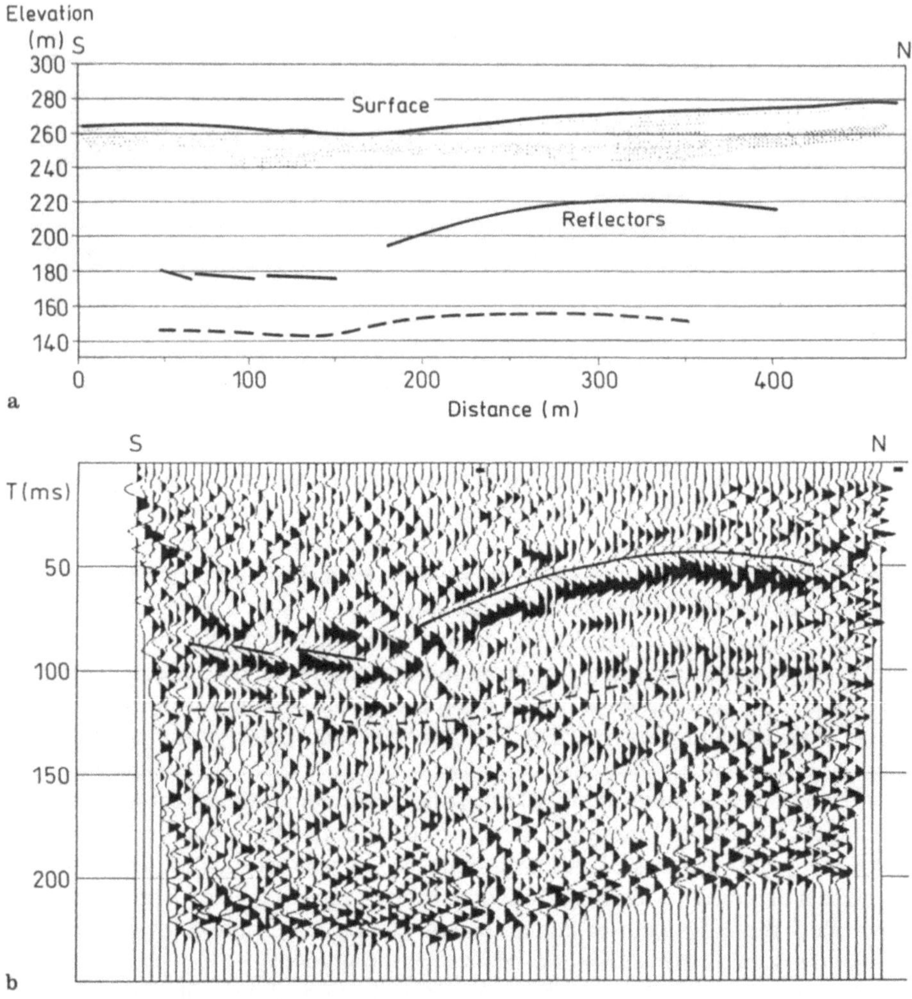

Fig. 3.53 a, b. Seismic reflection structures below a hazardous waste site. **a** seismogram of hydraulically active fault zone; **b** interpretation.

holes; it lies approximately between beds 1 and 2. Obviously, it has not acted as a reflector. The horizons of the refraction depict strong undulations of the boundaries 2/3. The corresponding reflector R1 occurs only in the left part and does not follow the ups and downs of refraction. However, the reflectors R2 (in part) and R3 (in total) could not be found by refraction: they lie deeper.

The combination of the two seismic methods makes it clear that both lead to different results. Whereas seismic refraction is well adapted to the structure built on and directly below waste deposits, reflection is better suited to unearth the structures of the deeper underground.

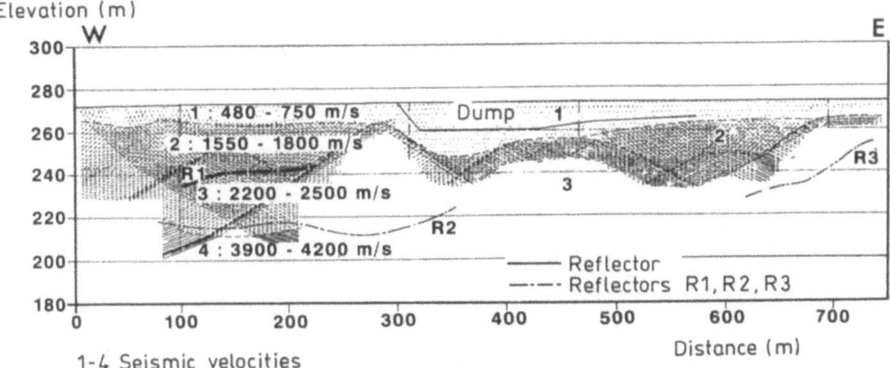

Fig. 3.54. Combination of seismic refraction and reflection at an industrial hazardous waste site

3.1.6 Gravity

Gravity surveys are rarely employed to solve environmental problems. The reasons for this are not only the relatively high cost, but also the difficulty of recognizing small bodies like buried waste of unknown density within the pattern of gravity anomalies. In addition, the necessary corrections depend very much on the relief of the vicinity and may obliterate the weak anomaly of a contaminated object.

Figure 3.55 presents a gravity survey of the hazardous waste site of Fig. 3.50 (seismic results). The stations were spaced 10 m along a 200-m long gravity section. The evaluation enclosed the free air and the Boguer corrections. After the free air correction, the waste deposit is still visible as a weak negative anomaly, but it diminishes completely after the Boguer correction has been applied.

Therefore, it is compulsary to correct raw gravity data completely by applying free air and Boguer corrections; otherwise, false interpretations can be worked out.

The landfill in Indiana, USA, whose geomagnetic survey is depicted in Fig. 3.8, was also investigated by a gravity survey. Figure 3.56 shows the complete Boguer gravity contours obtained from 200 gravity stations at intervals of 5–10 m at 8 lines over and adjacent to the landfill.

Figure 3.56 displays a regional variation of the area due to changes of the lithological facies of glacial sediments and bedrock. This regional trend was subtracted from the Boguer anomalies to yield the residual gravity contours of Fig. 3.57. The interpretation of the gravity data to elaborate the underground shape of the landfill was twofold:

1. Forward modelling with a constant density contrast of 0.53 g/cm^3 between the landfill and the glacial sediments led to the gravity section of Fig. 3.58.
2. Inversion of the residual gravity data by using knowledge of the actual dimensions of the landfill, based on varying density contrasts between sections (Fig. 3.59).

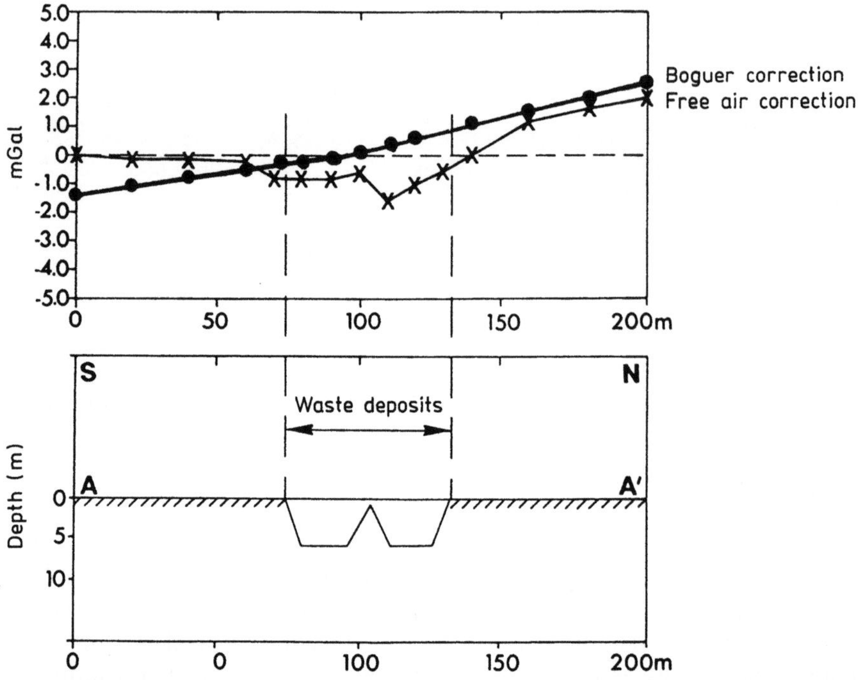

Fig. 3.55. Gravity survey of a hazardous waste deposit with free air (crosses) and Boguer (black circles) corrections

Apart from these good results, more physical properties of the dumped material could be deducted from the gravity data. A calculated increase in density contrast from the north to the south of the landfill is observed. It is attributed to stronger compaction of the older northern part and to a change in the composition of the fill material.

In the north, domestic trash of higher density predominates; in the south, construction refuse and brush cuttings of lower density are in the majority. Furthermore, a water saturation of 20% and a porosity range from 43% to 48% is anticipated from the computed density contrasts.

However, the gravity method may not be suitable for the investigation of all waste sites. This gravity survey was possible because it depended on a profound density contrast between homogeneous glacial sediments and fills.

3.1.7 Geothermy

Geothermal investigations depend on a strong heat flow on the surface of hazardous waste deposits. Unfortunately, this specific heat flow is often smaller than the heat produced by the rays of the sun. Rain and other meteoric influences compete with the sun and create heat flows that are difficult to assess. Therefore, geo-

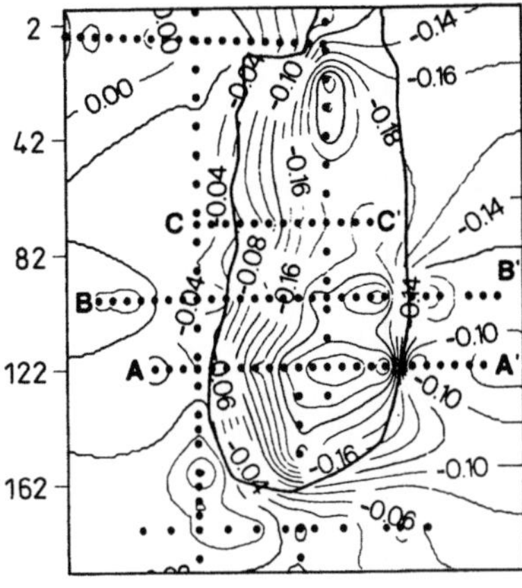

Fig. 3.56. Boguer gravity contour map of Indiana landfill

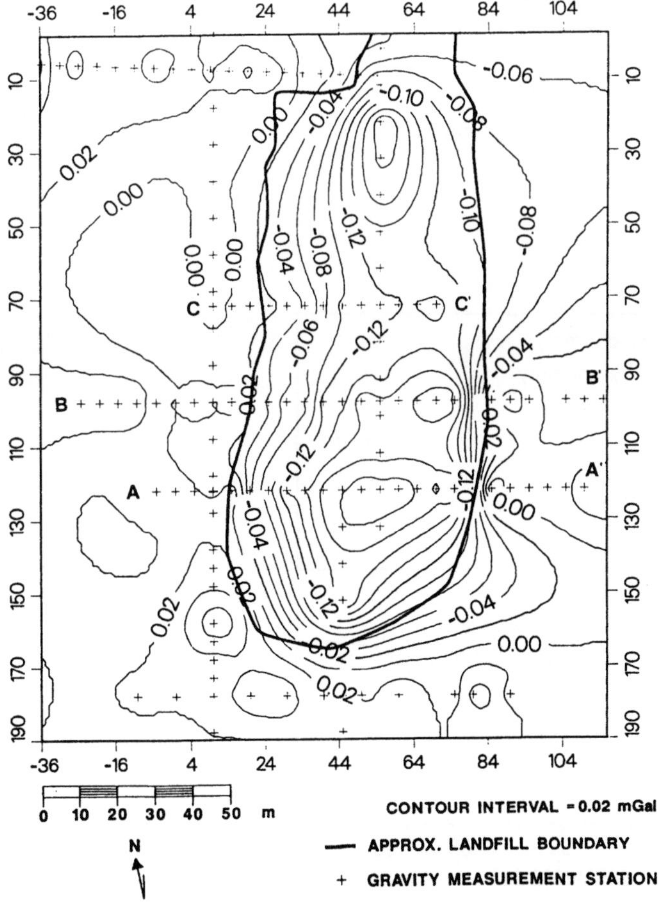

Fig. 3.57. Residual gravity contour map of Indiana landfill

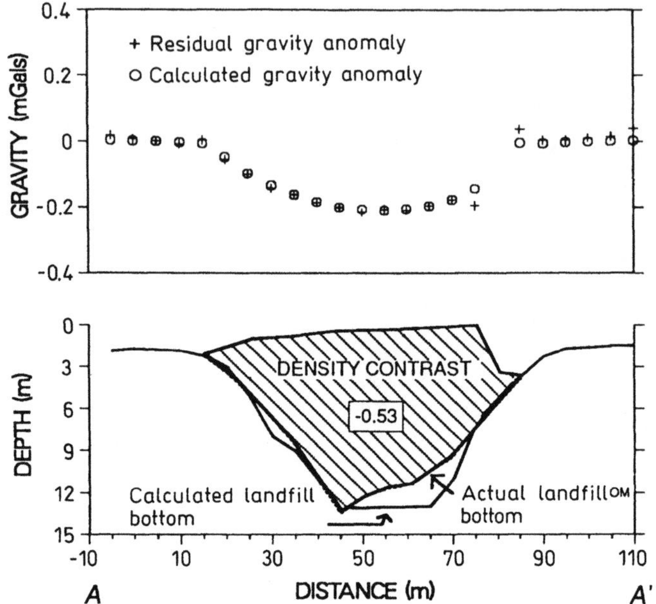

Fig. 3.58. Forward-modelled gravity data of Fig. 3.56

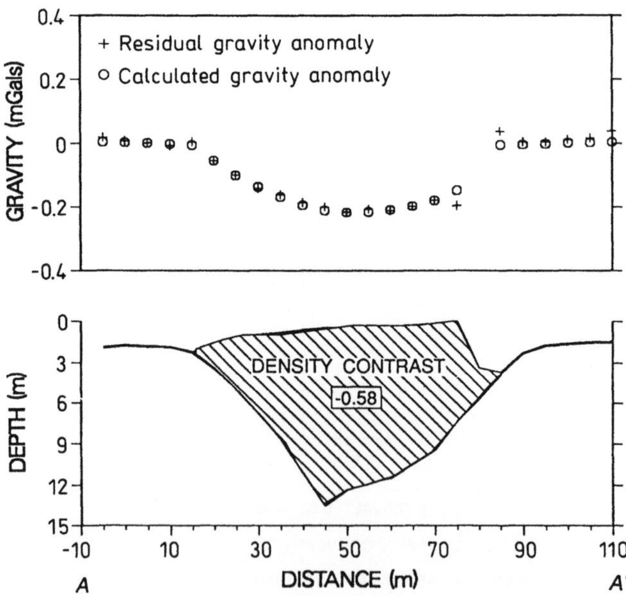

Fig. 3.59. Inversion of gravity data of Fig. 3.56

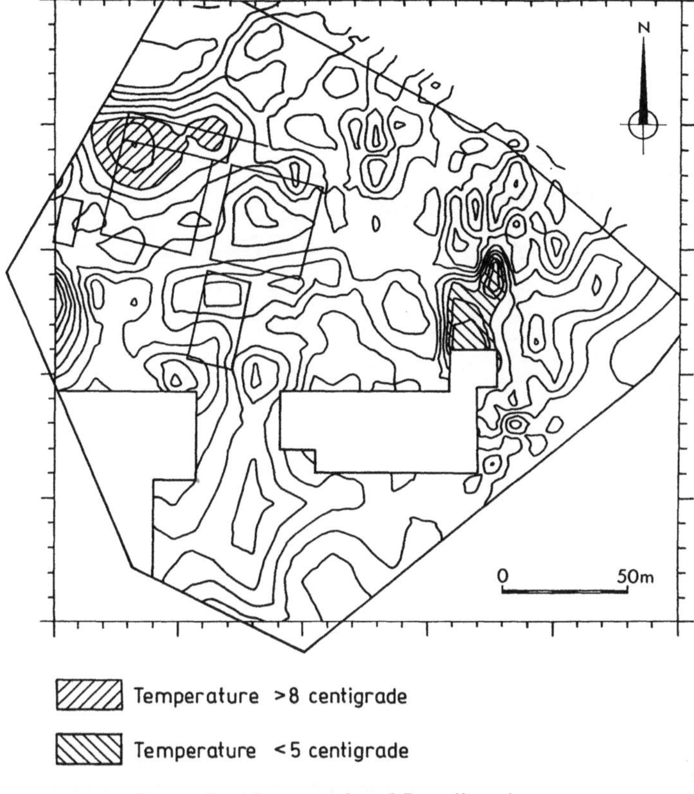

Fig 3.60. Geothermal contours of a buried hazardous waste deposit. The area is used by a tennis club

thermal measurements should be performed only in the middle of the night and in dry weather.

The temperature should not be determined right at the surface, but preferably in shallow probing holes of 0.5 – 1-m depth.

Figure 3.60 presents the outcome of such a survey. The temperatures, which were taken from probes at 0.5 m depth, appear to be nearly homogeneous over the waste deposit. Exceptions are a temperature increase in the area of the tennis courts, and a temperature decrease in the east near a steep slope. The positive anomaly stems from the absorption of the sun rays by the red clay of the tennis courts; the negative anomaly is correlated to seepage water on the surface.

Most of the geothermal indications of Fig. 3.60 are due to surface effects by sun rays or evaporation of water. Obviously, no heat production by oxidation or fermentation was traced on the surface. Since this overwhelming surface influence of the daily weather is seen in the thermal behavior of many waste deposits, thermal scanning from the ground or by airborne survey may fail.

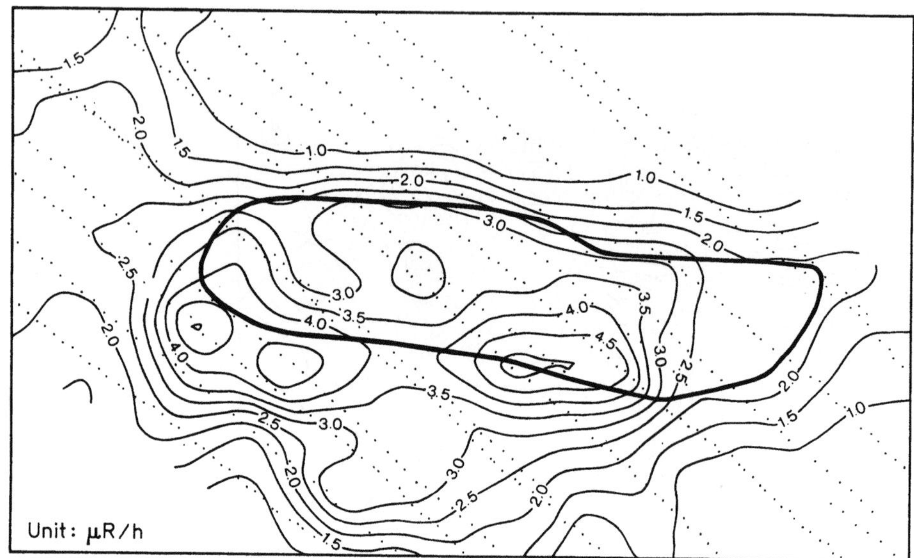

Fig. 3.61. Radiometric contours of the domestic waste dump of the city of Hannover, Germany (Figs. 2.9 and 3.47). Flight line spacing ~ 50 m (1 μJ/h = 3876 μR/h)

3.1.8 Radiometry

Hazardous or other waste dumps mostly show only weak radiation. The effects of geological or artificial clay barriers caused by the nuclear decay of the potassium isotope ^{40}K may be stronger than the radioactivity of domestic waste.

This becomes evident by a radiometric helicopter survey flown at a height of 50 m across the waste disposal dump of the city of Hannover, Germany (Fig. 3.61). Though the radiation dose increases from 10 μR/h in the vicinity to 50 μR/h at the center of the dump, this anomaly is not significant for domestic garbage because it could also have originated from other sources, as from the mentioned clay beds.

Stronger radiation comes from the industrial waste of uranium mines. Sources are the dumps of tailings and refuse from mining and ore dressing. In eastern Germany, large areas of the former Soviet uranium mining industry are contaminated by radioactive material. Remedial actions can only mitigate the impact of radiation because radioactivity cannot be destroyed either by chemical treatment or burning.

However, gamma radiation can be completely shielded by a 3-m thick sealing of clay. Figure 3.62 shows that such an overburden absorbs a radiation of >75 000 Bq (becquerel) completely.

On top of the dump of uranium tailings (Fig. 3.62), a clay seal of 3-m thickness conceals the gamma radioactivity completely. Over the uncovered slope, it increases, however, to >75 000 Bq; a smaller increase occurs over the river bed, where radioactive refuse has accumulated.

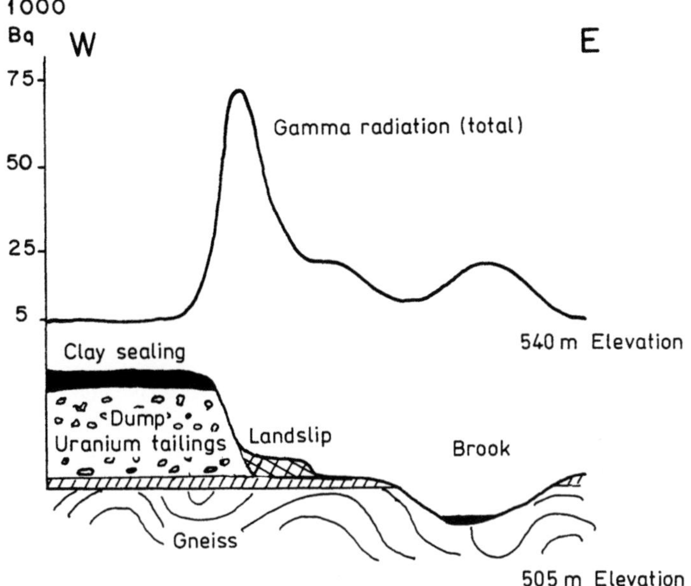

Fig. 3.62. Radiation survey of an uranium dump

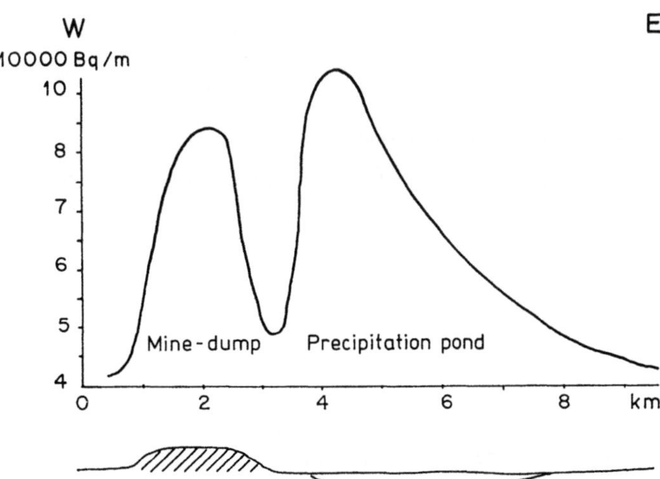

Fig. 3.63. α-Radiation in the ground air, equivalent to the content of radon gas. The section crosses a mine dump and a large precipitation pond. The ground air samples were drawn from soil in a depth of 0.9 m

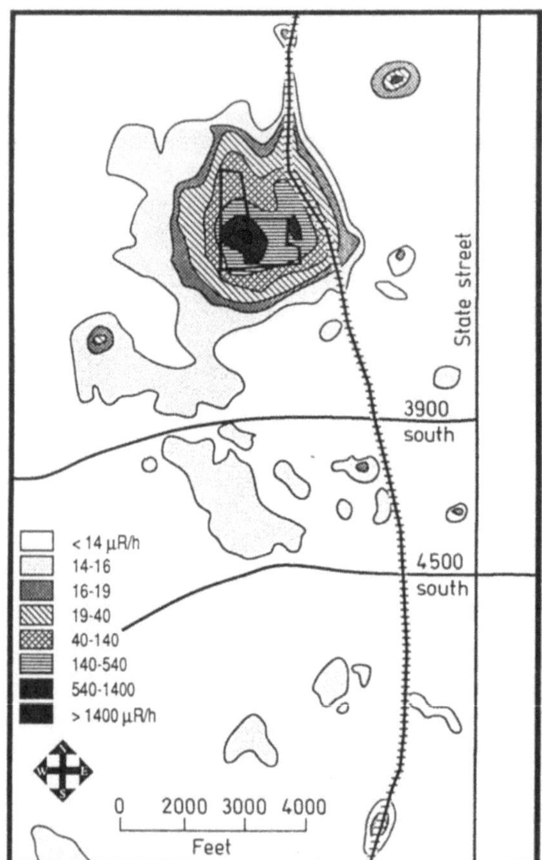

Fig. 3.64. Contours of radiation (total exposure rate) over vitro tailings (heavy outline) calculated from airborne radiometric data

While the direct impact of γ-radiation can easily be shielded, the gaseous nuclide radon (^{222}Rn) (Fig. 2.28) will nevertheless leak to the surface and into the air. An example of the determination of radon by the measurement of the α-radiation in the ground air is given in Fig. 3.63. In the area of Ronneburg/Germany, the ground air was sampled by a steel probe at a depth of 0.9 m. Though the 100-m spacing of the stations was coarse, it immediately gave a statement about the distribution of radon gas (Fig. 3.63).

The short half-life of 3.823 days demands quick progress of the survey and immediate analysis of ground- or pore-air samples in the field. By observing this rule in Ronneburg, a very high background of 45 000 Bq/m^3 was found. This was eight times higher than the regional background outside the area of uranium

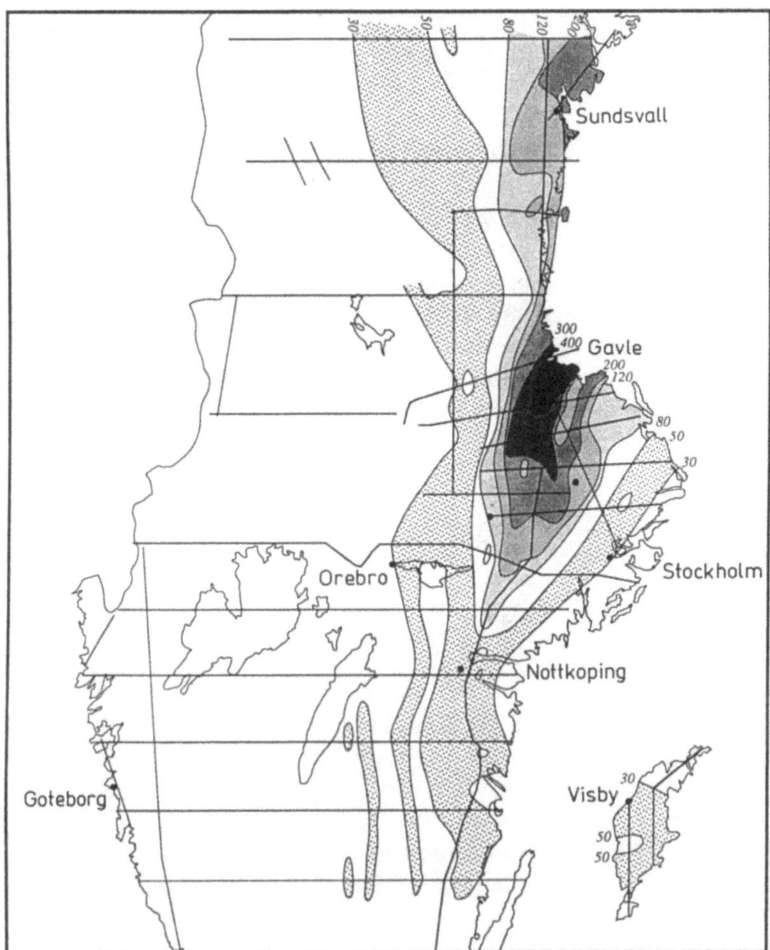

Fig. 3.65. Airborne gamma radiation map in mR/h after the Chernobyl fallout over southern Sweden, flown 1–6 May 1986

mining. The two radon peaks of 90 000 and 105 000 Bq/m^3 in Fig. 3.63 are caused by two mining structures: the mine dump and the precipitation pond.

Radon contaminations are not only caused by mining activities; the natural occurrence of uranium certainly produced radon long before the first miner appeared. But this amount remains mostly unknown, since it is rarely measured.

Another good example of a radiometric survey is provided by the vitro tailings site in Salt Lake City, Utah, USA. A helicopter flew a survey at 46-m altitude and 76-m line spacing. The scintillometer contained 20 crystals of NaI with a total volume of 12 914 cm^3.

Figure 3.64 presents the contours of the rate of total exposure over the vitro tailings site. Whereas the background radiation lies between 9 and 16 μR/h, the vitro tailing pile reaches a peak of >1400 μR/h. This immense gamma ray intensity results in a "shine" that extends 650 m away from the source. This effect has masked both windblown tailings and tailings that were used in a small parking lot nearby.

The lobes of higher radiation around and south of the tailings pile have different origins. Some are caused by windblow or by uranium ore lost from freight trains, industrial slags or milling equipment.

On 28 April 1986, the nuclear accident at Chernobyl/Ukraine resulted in radioactive fallout over parts of Sweden. A radiation air survey was flown at a height of 150 m. Its spectrometer measurements resulted in the gamma radiation map of Fig. 3.65 for the southern part of Sweden.

The last two case histories prove the importance of radiometric measurements for the detection of extremely dangerous contaminations which cannot be traced by human senses.

3.2 New Disposal Sites

3.2.1 General

Whereas the investigation of hazardous waste sites is a new task for geophysicists, the exploration of new disposal sites can rely on ample knowledge and experience. This has been assembled by numerous geophysical surveys for ground water. The problems are equivalent: localizing active hydraulic features means also finding hydraulically paths of leachates or seepages. Such hydraulically active structures or aquifers in unconsolidated rock, such as sandor gravel beds, and in hard rock, such as fissureor fault zones, must be avoided.

3.2.2 Horizontal Stratification

Vertical Electric Sounding (VES)

For five decades, this method has proven successful in securing the resources of ground water. It is also well adapted to clarify the hydraulic permeabilities of strata in the vicinity of existing waste dumps or in areas where new waste sites are planned.

Its main advantage is the precise distinction it makes between aquifers in unconsolidated rocks like gravel and sand with high resistivities and clay-bearing aquicludes with low resistivities.

Every VES results in a column enclosing the apparent specific resistivities in Ωm and the depth of boundaries between layers with different resistivities. VES can therefore be compared with a borehole. The resistivities inform about the permeability of the beds; the depth or thickness of the strata tell about the volume of hydraulic or geologic features. Despite this advantage, VES costs only a few percent of the cost of a borehole and is much faster in producing results.

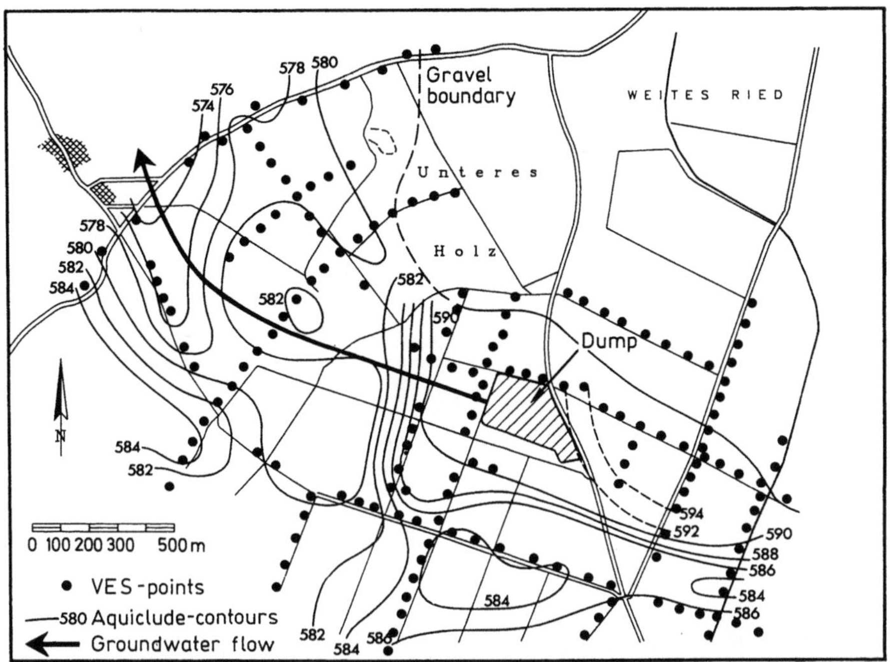

Fig. 3.66. Contours of an aquifer base in the alpine foreland by VES. A contaminated plume follows an underground channel, carved into the top of the glacial till (arrow). VES electrodes were separated up to 160 m; 9 profiles with a total length of 17 km were investigated

However, some boreholes are always necessary to verify the VES data. VES at the location of boreholes allow a precise calibration of the sounding curves to the hydraulic permeabilities of strata or of tectonic structures.

Figure 3.66 presents VES results in the vicinity of a waste dump in a plain in the alpine foreland. The contours show the base of the aquifer in meters above sea level. The resistivity of the gravel aquifer varies between 700 and 2500 Ωm. The aquiclude below consists of glacial till and drift with lower resistivities from 20 to 150 Ωm. The boundary or the base of the aquifer is thus marked by a distinct change of resistivity and can easily be followed up by VES.

This survey resulted in a detailed relief map of the aquiclude, with its surface structures guiding the flow of ground water and of a contaminated plume. In this case (Fig. 3.66), the plume was traced, though it was only weakly saline and of similar resistivity as the uncontaminated aquifer. It was located indirectly by mapping the underground structures guiding its flow.

The hydraulic impermeability of the geological barrier is the most important precondition for the construction of new waste disposal sites. The next case (Fig. 3.67) describes the search for such an impermeable area in a plain of glacial drift. A grid of 1600×600 m^2 with a rectangular mesh of 100×100 m^2 was surveyed by 96 VES.

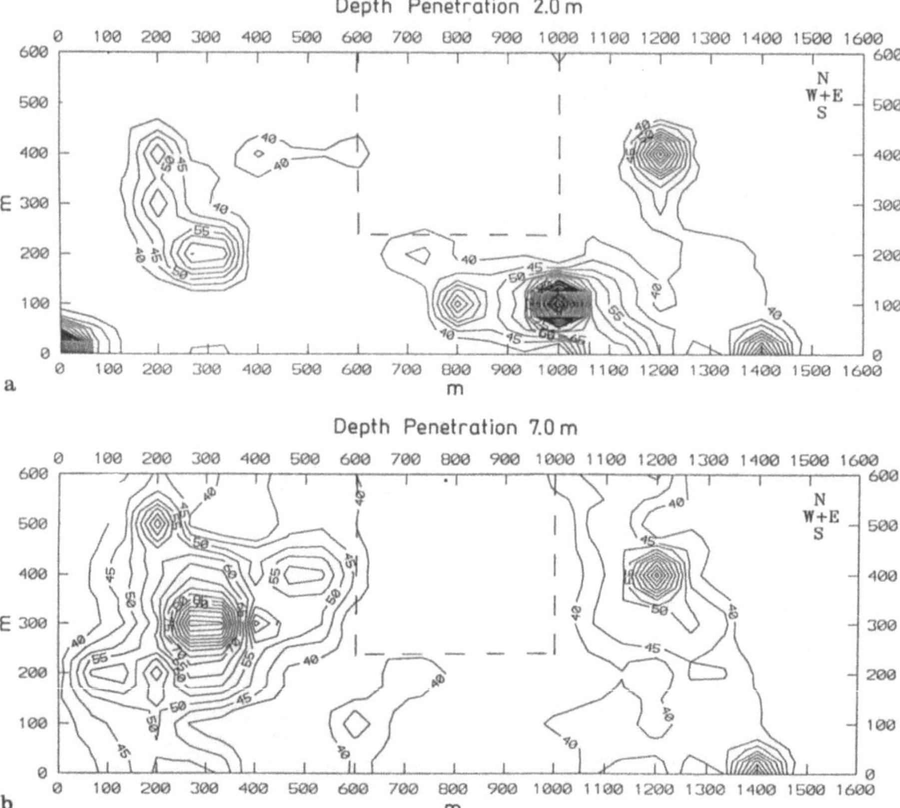

Fig. 3.67 a, b. Contours of the apparent specific resistivity at different depths, derived from 96 VES. A possible new dump area is framed by dashed lines. It consists of impermeable rocks with resistivities < 40 Ωm, indicating a good geological barrier

The target was to eliminate all areas with permeable soil and rock. From the VES curves, two resisitivity contour maps for the depths of 2 m and 7 m were derived. In Fig. 3.67, resistivity contours were drawn only from 40 Ωm upward, showing that most of the area contains rocks with higher resistivities and thus too-high permeabilities. Only the rectangular area in the center, framed by dashed lines, displays resisitivities < 40 Ωm. Since a countercheck by drilling found impermeable clay in the underground, this area was finally chosen as a new waste disposal site.

This geophysical result reduced the area to be investigated by drilling or probing from 960 000 m^2 to 123 000 m^2! That means the development costs were reduced by approximately 87%. Thus, considerable savings were achieved by spending only US $ 12 000 for the geoelectric survey.

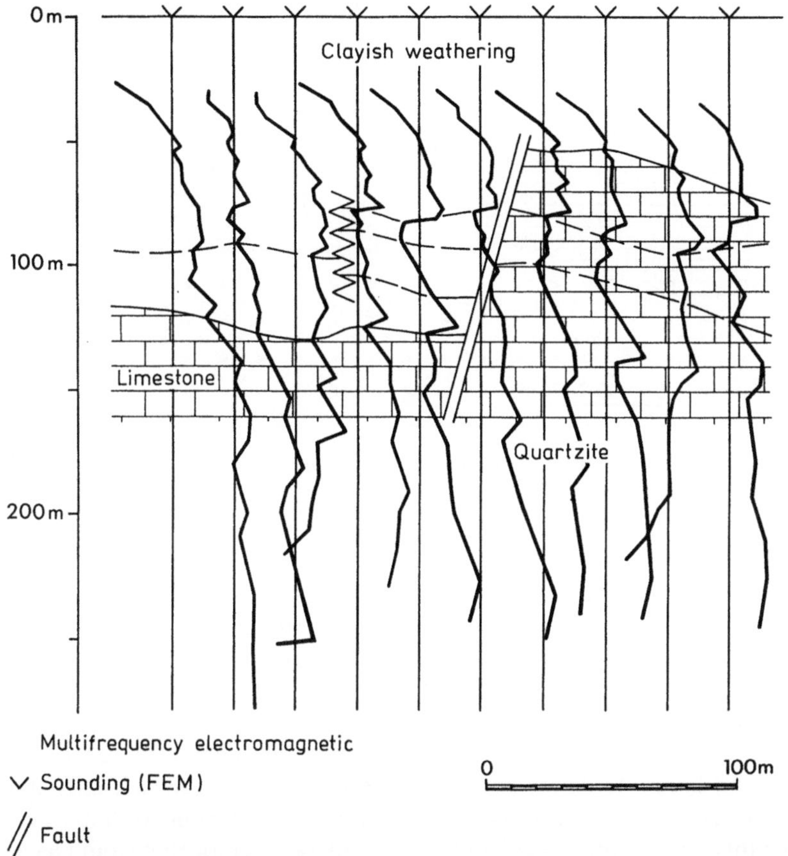

Fig. 3.68. Cross section of 10 FEM soundings. Thick layers of limestone and quartzite are over-burdened by weathered clay and loam. A fault thrusts the left block > 50 m down. This interpretation was confirmed by follow-up drilling

Electromagnetic Sounding

This method is also known as multifrequency sounding, or FEM. It was developed in Canada and in the Geophysical Institut ELGI, Budapest, Hungary (see also Sect. 2.2.2). Figure 3.68 presents an FEM section of stratified limestone overlain by weathered clay. A quartzite was found deep under the limestone. The sudden change of the limestone thickness is believed to be caused by a fault down-thrusting the northwestern block.

Seismic Refraction

In the vicinity of a hazardous waste dump, a line of 410 m length was surveyed by seismic refraction. The geophone separation was 5 m. Figure 3.69 shows the

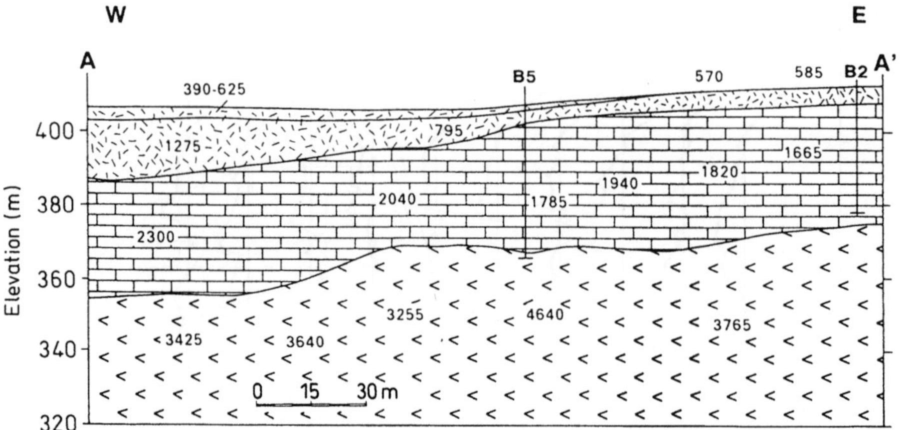

Fig. 3.69. Geological model by seismic refraction. Thickness and dip of the sandstone (1665 – 2300 m/s) were confirmed by drilling

achieved structural model. Under a low-velocity overburden lies a sandstone bed of ~ 40 m thickness with seismic velocities varying from 1665 to 2300 m/s. This was interpreted as the indication of an inhomogenously fissured sandstone. By extending the spread of geophones to 240 m, another, deeper bed of selenitic Keuper with > 3000 m/s was detected.

Follow-up drilling confirmed this seismic prediction. The depth difference between seismic and borehole data was less than 2 m. By refraction, not only the thickness of the sandstone but also its dip to the west was determined. Important is the fact that this seismic profile coincides with a metal gas pipe that would have obliterated all geoelectric signals from the ground.

Seismic Reflection

Its advantage lies in the possibility of recording many reflecting horizons at once and of clarifying even complicated features of the underground. Its disadvantage is its inability to record subsurface structures.

Figure 3.70 describes a digitally evaluated seismic section and its geologic interpretation of tertiary lignite seams, compressed by glacial pressure. The interruption lines of the reflecting horizons dip steeply down to 170 m. At greater depths, their dip flattens until an undisturbed series is reached in ~ 200 m.

Progressing glaciers have broken up the tertiary lignite series into separate blocks, piled them up and created a complicated geohydraulic system with steep sutures. Such an area is definitely not suitable for the construction of new waste disposal sites, even if the subsurface beds consist of impermeable clay, because the weight of the deposited material could release landslides and slumps.

In the following example of seismic reflection (Fig. 3.71), subsurface features were detected by determining the "optimum reflection offset window." This is the window that allows shallow reflectors to be observed with a minimum of seismic

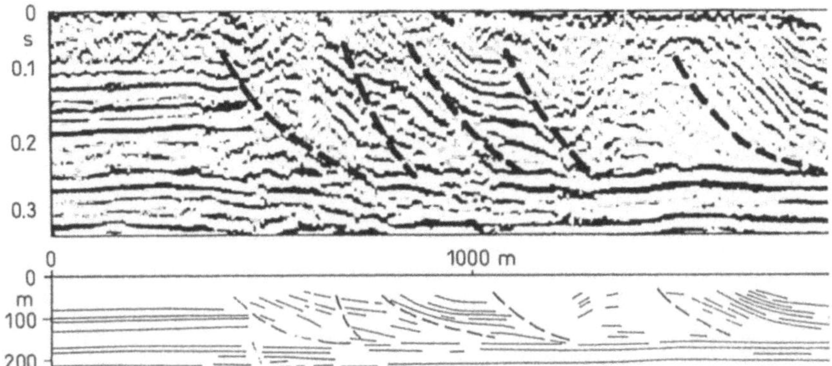

Fig. 3.70. Cross section of seismic reflection in tertiary strata. Glacier-pressure has piled up lignite beds

noise. It has to be found by trial and error, e.g. by shooting several expanding spreads around the area.

Figure 3.70 presents a seismic section with derived geology. Flat and dipping reflectors of sediments disclose the piling-up of lignite beds. This renders this area unsuitable for the construction of a new disposal site.

Figure 3.71 introduces a sophisticated interpretation of a seismic reflection survey. Three continuous reflectors between 20- and 50-m depth indicate flat-lying glacial lake sediments. Such stable and unfaulted subsurface conditions stand for low ground water risks and mark the site as a potential hazardous waste treatment facility.

3.2.3 Steep Dipping Structures

Geomagnetics

Geomagnetic measurements can sensibly be used not only to trace buried magnetic waste but also to find natural magnetic structures that influence the path of contamination. In Fig. 3.72, a dyke of basalt acts as a hydraulic swallow path for contaminated surface water.

From the foot of a waste dump, the leachate flows over the surface until it seeps into a cleaved and fissured basalt dyke. This guides the contamination through an impermeable clay barrier deep into the ground. The dyke was detected by a geomagnetic survey under thin loamy overburden. The inclination of the magnetic earth field of 55° causes only the southern maximum to coincide with the top of the dyke. The related minimum lies north of this magmatic and magnetic structure.

Electromagnetic Mapping

This method is apt to locate steep dipping and long-extending hydraulic structures in karst and hard-rock areas rapidly and at low cost. Due to its inductive coupling,

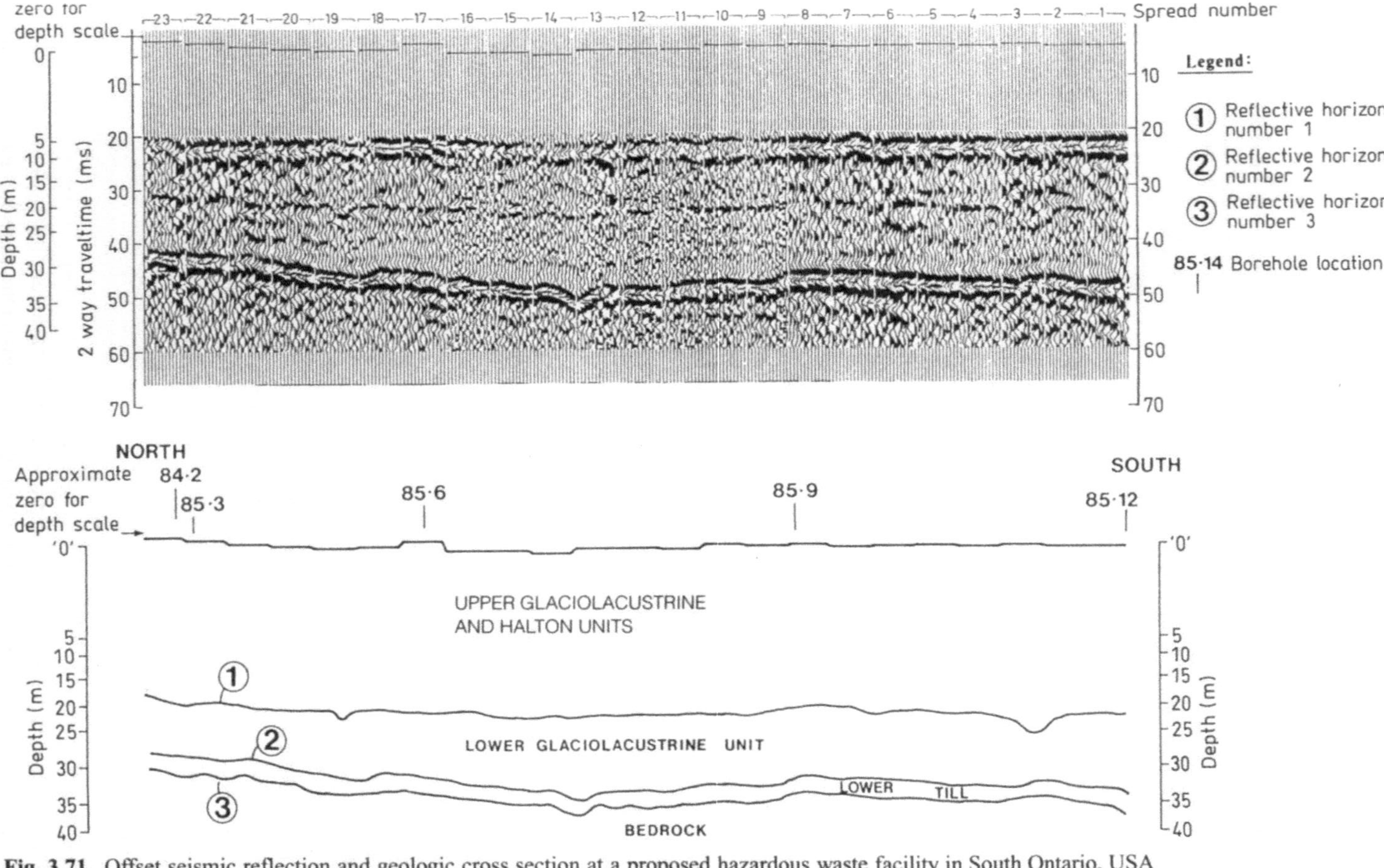

Fig. 3.71. Offset seismic reflection and geologic cross section at a proposed hazardous waste facility in South Ontario, USA

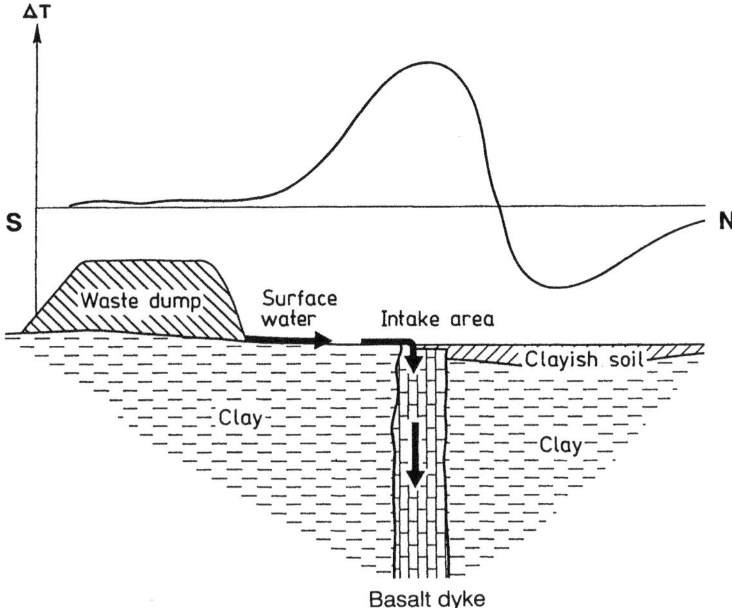

Fig. 3.72. Cross section of magnetic total intensity ΔT. The seepage path of a basalt dyke penetrates a clay barrier

no electrodes have to be driven into the ground. Further advantages are the high resolution of steep sheet structures like fault planes, fissured zones or crevices. The linear extension of such structures is won by correlating the related EM indications of several profiles to "linears".

Linears are marked by pronounced minima of the in- and outphases of the electromagnetic field. Needless to say, the EM-profiles should cross long tectonic structures perpendicular to their strike to achieve undistorted values and to avoid errors of interpretation.

Each EM survey point must be entered precisely into the location map. Numbered pegging is not necessary if the EM profile is a straight line, terminated at both sides by vanishing poles. The point separation can be drawn from fixed marks at the connecting cable between transmitter and receiver.

EM results always pertain to the midpoint of the EM spread. Since this is sometimes overlooked, a countercheck and perhaps an appropriate correction is advisable.

Figure 3.73 portrays an electromagnetic section across a planned new waste disposal site. Minima that are typical for well-conducting linears occur with changing intensities and shapes over faults and fracture zones. While vertical fault planes cause symmetric minima right above the structure (A and E), inclination accounts for a shift of the minimum towards the side of the dip (B). Structures where in- and outphase curves run in opposite directions are attributed to subsurface effects.

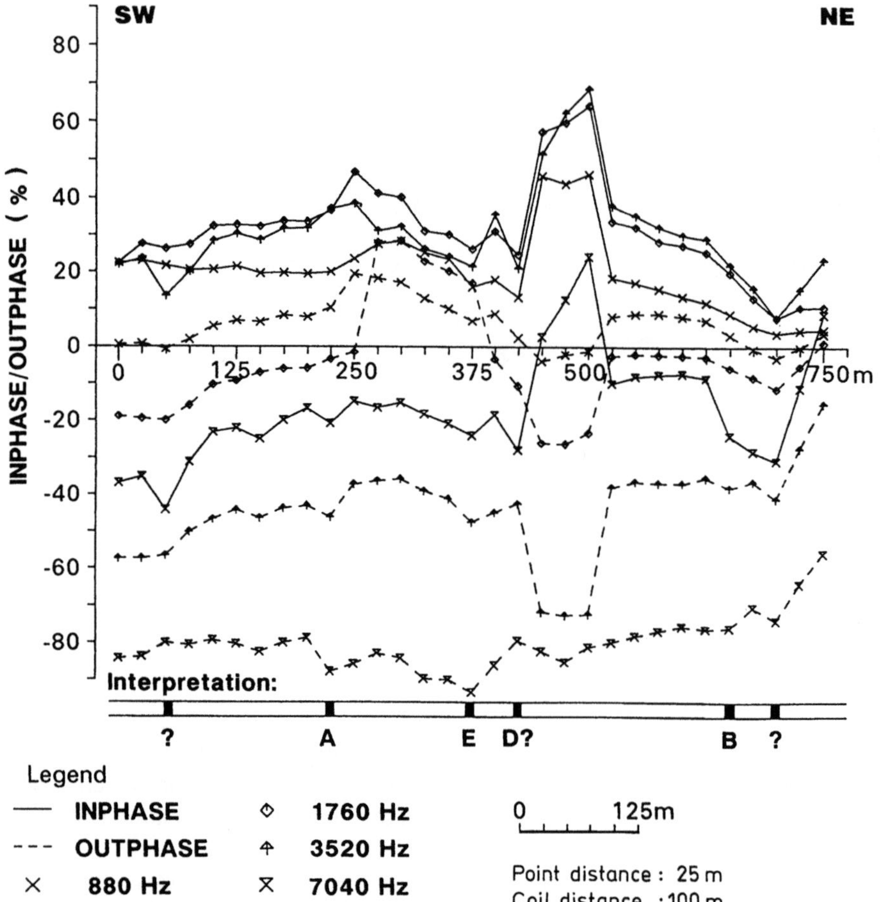

Fig. 3.73. EM cross section of a planned waste site. Minima carry capital letters and connect up to long linears (Fig. 3.74)

If minima move too close together, their interpretation may be impounded. Uncertain minima are question marked in Fig. 3.73. The construction of linears between the minima of different cross sections always requires close cooperation between geophysicists and geologists, geohydrologists or environmental experts. A geophysicist alone, without relevant knowledge, would perhaps draw pointless structural patterns.

The EM contour map of Fig. 3.74 provides a good example for such a cooperation. The linears are derived from the minima A to F of the cross section Fig. 3.73. It is worth noting that linears cannot be constructed from contour lines. The geophysicist must consider the shape of the minima in every case and has to consider appropriate model structures to achieve a proper interpretation.

The linears of this area were identified by a drilling programme as an echelon of faults with vertical and horizontal thrusting. They do not render the area

Fig. 3.74. EM-contours and linears (black beams), constructed from EM cross sections (Fig. 3.73). Inphase data, frequency 3520 Hz, spread 100 m, + = measuring point, separation 25 m

permeable and unsuited for waste dumping, however, since the weathered fault planes are sealed by clay and loam. These faults do not guide but retain the ground water as steep dipping aquicludes.

The most appropriate frequencies for electromagnetic mapping of such steep dipping zones lie between 200 and 14 000 Hz. They guarantee good depth penetration by portable energy sources. Permanent distant transmitters with higher frequencies from 12000 to 24000 Hz, known as VLF, can also be used if the

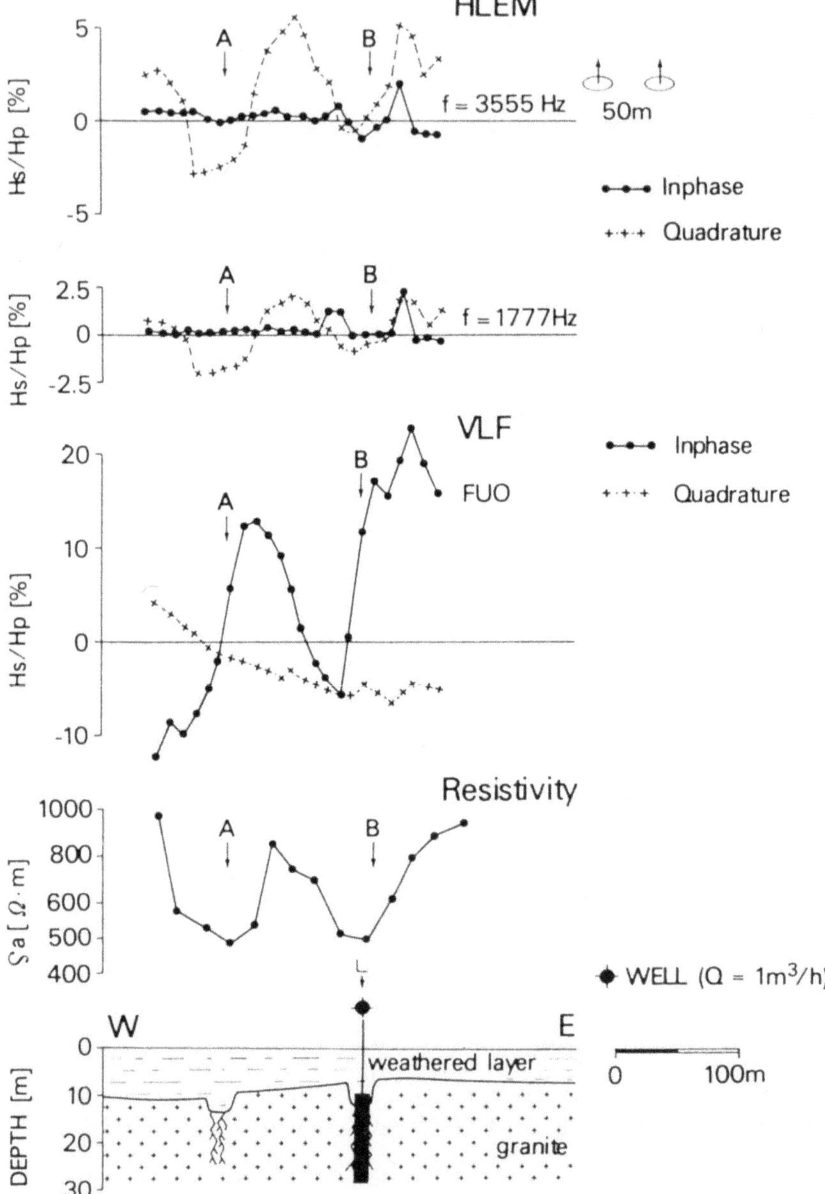

Fig. 3.75. Comparison of VLF and other electromagnetic and electric data over granite and volcano sedimentary rocks in Burkina Faso, Africa

objects are not too deep (Fig. 2.12). The results of VLF surveys are otherwise similar to those of EM mapping, as shown by Fig. 3.75.

The steep-dipping fracture zones of Fig. 3.75 are detected by all employed methods. Especially the VLF cross section responds well to the steep conductor, because it occurs inside the VLF depth limit.

Ground Radar

This method should be applied to the investigation of new disposal sites only if high resistivities or low dielectric constants allow the necessary depth penetration of >10 m. It can be used better to trace old underground mine workings or non-metallic pipes in prospective areas (Fig. 3.46).

Seismics

Up to now, refraction and reflection have been directed mostly at more or less horizontal structures. Steep dipping structures like faults had to be deducted indirectly from interruptions and dislocations of reflecting horizons (Fig. 3.53). The new 3D seismic method overcomes this handicap but it is perhaps still too expensive to follow up vertical textures, which might discharge contamination through hard rock areas.

Well Logging

Geophysical surveys at proposed new waste deposits should always be amended by geophysical logging, because the suitability of an area for the construction of a waste site depends not only on the porosity and permeability of the underlying hard rock, but also on the intensity and distribution of fracturing, which considerably effects its hydraulic properties. Geophysical logs reveal important additional information, even in cored boreholes.

In Fig. 3.76, seven salinometer logs explicitly exhibit the saline-water distribution, mixing zones and freshwater occurrences in seven pumped wells. Water inflows were located by differential temperature logs. Such results are of great importance for the control of freshwater resources in areas where fresh and saline ground water occur together. The freshwater resources that are found in lenses under many islands need to be checked by such logs regularly.

Since most fractured zones are difficult to drill and produce poor core recovery, geophysical well logging is essential in such boreholes. In Fig. 3.77, not only the caliper log shows fracture zones as backbreaks of the borehole wall; they are even better recorded in the sonic and electric logs.

Figure 3.77 displays geophysical logs in gneisses combined from 13 different runs and methods. The uniform logs of the electric resistivity, the gamma radiation, the sound (seismic) velocity, and the caliper (borehole diameter) indicate a fairly homogenous petrography with constant ground water permeability. Solely the zones with sudden high chargeabilies of induced polarization mark fractured hydraulic paths for the movement of ground water and soluble pollutants.

Another advantage of logging is the verification of geophysical statements about the depth of structures. This corroboration is necessary for depth data of

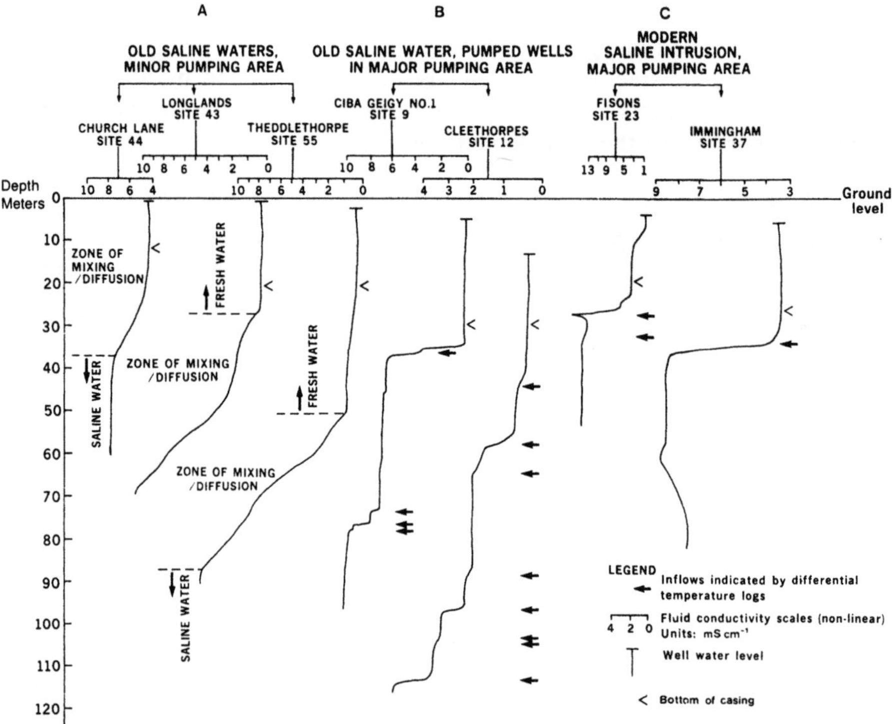

Fig. 3.76. Geophysical salinometer logs, Lakeview, Ontario, Canada. Arrows indicate inflows found by temperature logs

geoelectric soundings (VES) and seismic refraction. Geophysical logging can furthermore be employed to revise the depth figures given by the drilling crew.

The petrographic description of the drill cores of Fig. 3.77 contains no clues as to the nature of those zones. Neither rock boundaries nor other textures were described. Only the microscopic investigation revealed disseminated graphite and sulfides grown inside zones of tectonic shear and thrust.

This example proves the high value of well logging, since zones of tectonic weakness in hard rock, which may guide the spreading of contaminants, can be located with more precision than by the petrographic analysis of drill cores.

Radiometry

An increase in the radioactive gas radon in the ground air may point to steep dipping and deep-reaching structures. The contours of α-counts in Fig. 3.78 display variations in the radon content above the normal background. The background was established at 150 α-counts (cpm) in a nearby well.

The investigated area houses a hazardous waste deposit of chemical refuse. West of the dump, α-counts as high as 450 cpm, accompanied by high helium contents of the ground air, were measured. This double anomaly was interpreted

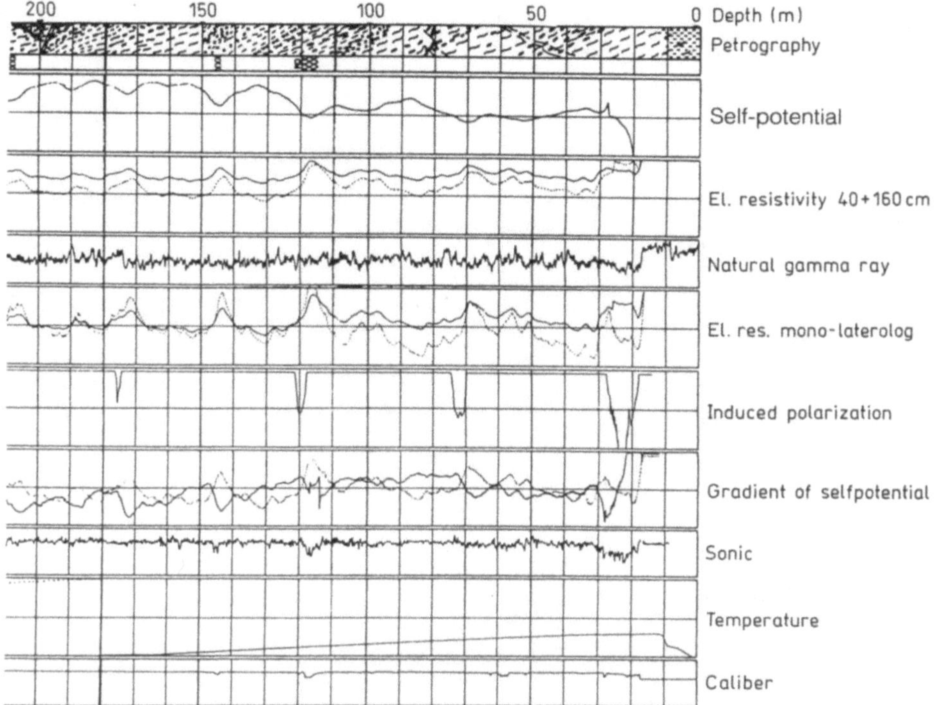

Fig. 3.77. 13 Geophysical logs from a borehole in gneisses in Bavaria, Germany

as an indication of a deep-reaching fracture zone allowing the radon gases to ascend to the surface. An origin of radon from the waste deposit could be excluded.

3.2.4 Nuclear Repositories

It is imperative that, in the future, nuclear waste is kept out of the biosphere or the ground water system. Appropriate rocks are dry granite or gneiss and, most important of all, salt domes. Salt is absolutely dry and encases hot, highly radioactive waste completely because of its enhanced plasticity under high temperature and rock pressure.

No alien rocks or even lye are allowed in a nuclear deposit, because such pockets might expand when heated and open up cracks or fissures for the intrusion of ground water. They can certainly be detected by drilling. However, finding all inclusions would necessitate many boreholes and convert the salt dome not into a nuclear deposit, but into a Swiss cheese.

To overcome this and to investigate salt diapirs spatially, the method of electromagnetic reflection (EMR) or georadar has been developed. EMR is able to

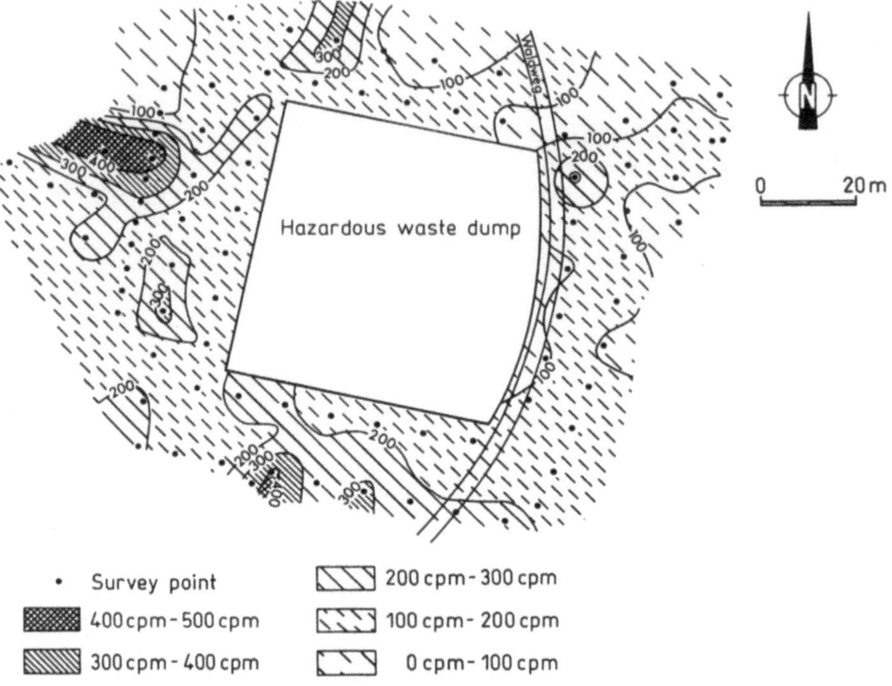

Fig. 3.78. Radon contours in the ground air, measured as α-counts per minute near a hazardous deposit of chemical waste

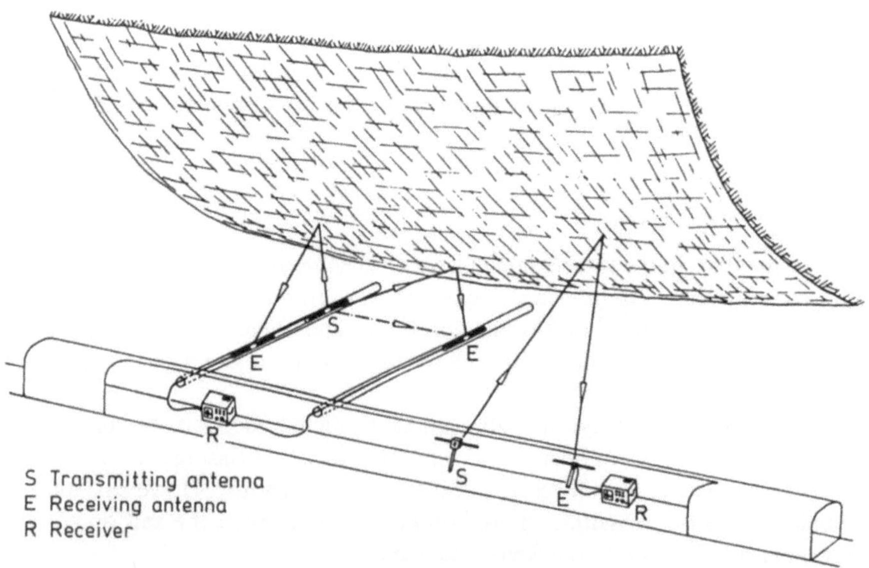

Fig. 3.79. Locating a thin reflecting clay horizon by EMR in a salt dome from two horizontal boreholes and from an adit

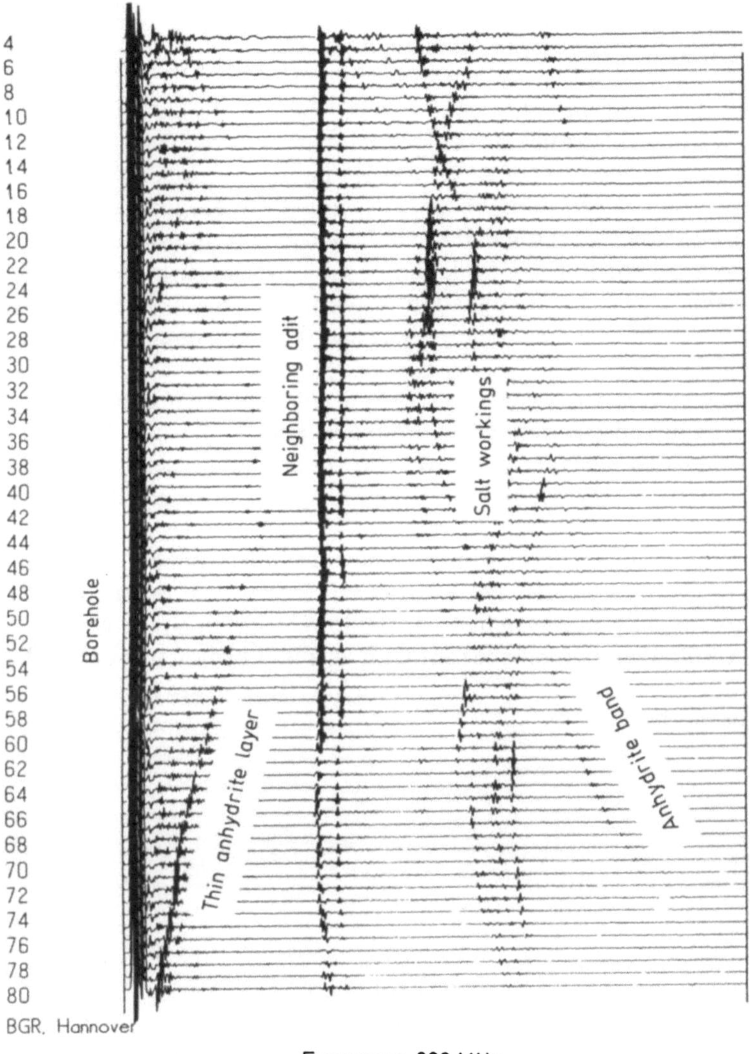

Fig. 3.80. Underground radargram measured in a salt dome to be developed as a nuclear waste repository

locate any alien inclusion precisely without endangering the safety or the sealing of the future disposal mine.

EMR works with the same high frequencies (>100 MHz) as the ground radar. However, transmitting and receiving antennas are separated. This optimizes the signal reflection and allows determination of the direction of the EMR reflector.

The extremely high resistivity of the salt allows the high-frequency EMR oscillations to penetrate and be reflected from distances between 5 and 1000 m. EMR is used to find the boundaries of different salt beds, anhydrite seals, clay lenses, lye-filled fissures, and intruded dykes of magmatic rock.

From underground adits or boreholes, the spatial position of all electric reflecting bodies with deviating dielectric constants can be established. Figure 3.79 shows how the bearings of an underground reflector are won by EMR.

According to the derived distance, very short pulses of groups of electromagnetic waves of only 0.1 μs duration are transmitted. This signal reaches the receiver after having travelled through the salt. It is also used to synchronize the receiver with the transmitter. The calculation of the travelled distance is made, according to seismics, by the determination of the travel time differences. But with EMR, the travel times are extremely short and their measurement requires great technological efforts.

The electromagnetic radar waves are received not only from one direction but from all spatial directions. Reflections from above the adit cannot be distinguished from those that stem from below. The spatial assignment of the reflections can be made either according to their shape or by using directional antennae.

In the radargram of Fig. 3.80 the two parallel reflections, running across the whole area, are caused by a parallel adit. Important information is revealed by the reflection curving slightly to the right from the lower margin (track No. 80 to track No. 40). A follow-up borehole identified it as a thin bed of anhydrite. The slightly bent reflections on the right beyond the adit reflection are the reverberations of old mine works.

4 Cost of Geophysical Surveys

4.1 Cost Structures

The expenses for geophysical investigations comprise the following:

1. Transport to and from the area of investigation,
2. Measurements in the field,
3. Evaluation and interpretation,
4. Reporting,
5. Overheads.

Tables 4.1 and 4.2 give a rough estimate of 1993 expenses for geophysical surveys. The prices include measurements, evaluation, reporting and overheads. Transport costs are not included. The given prices are non-binding, without obligations and guarantees as to correctness. They are meant to allow a quick estimate of geophysical costs. Special and local conditions may lead to higher or lower expenses.

In addition prices may be influenced by the following:

1. Volume of order; measurements become cheaper with growing volume.
2. Rough terrain increases, flat ground decreases the costs.
3. Imprecise or disturbed raw field data must be corrected by special digital evaluation programs at extra cost.
4. Many bidders may lead to reduction in costs.

Table 4.1. Estimated costs for one geophysical field day (1993)

Method	Cost per day (US\$)
Geomagnetics	500 to 700
Geoelectric mapping	530 to 1000
Geoelectric sounding (according to spread)	560 to 1100
Induced polarization	950 to 1300
Self-potential (>200 electrodes $\rightarrow 1000\,\$$)	350 to 1400
Electromagnetic mapping + VLF, TDEM	510 to 850
Georadar	900 to 1500
Seismic refraction	1000 to 2200
Seismic reflection	2000 to 10000

Table 4.2. Estimated costs of line meter and station (US$)

Method	Cost/m	Cost/station
Geomagnetic	1.50	1.50 – 3.50
Geoelectric mapping	1.45	1.45 – 3.00
Geoelectric sounding	1.65	42.00 – 120.00
Induced polarization	2.10	40.00 – 150.00
Self-potential	0.80	2.00 – 6.00
Electrom. mapping +VLF	1.45	1.45 – 3.00
Georadar	3.42	continuous
Seismic refraction	5.00	50.00 – 200.00
Seismic reflection	12.00	200.00 – 500.00

4.2 Comparison of Geophysical Expenditures

If a site has to be drilled or geophysically surveyed in a dense meshed grid, the following topics have to be addressed:

1. Extension of dumped waste or contamination,
2. Thickness of waste deposit,
3. Location of small (toxic) bodies.

Table 4.3 describes the actual cost of the investigation of a hazardous waste deposit. Under A) the expenditures without geophysics and under B) the costs with the application of geophysics are listed. The prices for geophysical measurements agree roughly with the figures given in Tables 4.1 and 4.2. Drilling was offered at US $ 150.00/m, shallow probing at US $ 30.00/m.

The cost-benefit analysis of Table 4.3 shows that the application of geophysics, plus a reduced drilling/probing program, has reduced the total cost of investigation by 73 %. The expenditures under A) were only planned; those under B) have been incurred.

A side effect of geophysics was the continuous spatial coverage of the waste site. Mere drilling and probing would have left gaps 5 – 10 m wide between the points. One can certainly imagine how many contaminated bodies, like poisonous sludges or bundles of drums with toxic filling, could have hidden between the drill sites.

The costs in Fig. 4.1 were taken from actual projects. They are, however, not representative for all investigations of landfills or waste deposits. Local conditions may vary considerably and influence the cost frame. However, it is a rule that the application of geophysical methods leads to considerable cost reductions.

In the case of Table 4.3, the cost decrease is 73 %. Any cost-benefit analysis of a landfill investigation must therefore include the reduction of the high costs of systematic drilling and probing by geophysics.

Table 4.3. Comparison of environmental investigation costs without and with geophysics. (Buried hazardous waste deposit covering $\sim 30\,000$ m^2)

A) Investigation costs without geophysics

330 Percussion probes, depth 6 m, grid 6×6 m^2
cost per meter US $ 20.00 .... US $ 39 600.00

18 cored boreholes, depth 25 m, grid 25×25 m^2
cost per meter US $ 150.00 .... US $ 67 500.00

Total expenditure: .... US $ 107 100.00

B) Investigation costs with geophysics

Magnetics, 1200 points 5×5 m^2 grid,
cost per point US $ 3.00 .... US $ 3 600.00

Geoelectric mapping, 300 points, 10×10 m^2 grid,
cost per point US $ 3.00 .... US $ 900.00

Geoelectric sounding, 48 stations, $L = 100$ m,
cost per sounding US $ 120.00 .... US $ 5 760.00

Seismic refraction, 500 line meter,
cost per meter US $ 8.00 .... US $ 4 000.00

30 Percussion probing, depth 6 m,
cost per meter US $ 20.00 .... US $ 3 600.00

3 Boreholes, cored, depth 25 m,
cost per meter US $ 150.00 .... US $ 11 250.00

Total expenditure: .... US $ 29 110.00

Table 4.4. Volume of geophysical measurements for the costs of one borehole meter

Method	No. of points	Line meters	Area (m^2)[a]
Magnetic	40 – 250	100 – 300	200 – 450
Geoelectric mapping	50 – 100	100 – 250	200 – 400
Geoelectric sounding	2 – 4	10 – 30	20 – 100
Induced polarization	1 – 4	20 – 30	50 – 100
Self-potential	25 – 75	100 – 300	200 – 500
EM mapping VLF, TEM	50 – 100	100 – 250	200 – 600
Ground radar	continuous	100 – 150	50 – 70
Seismic refraction	1 – 3	30 – 50	50 – 80
Seismic reflection	1 – 0.3	12 – 15	30 – 50

[a] Calculated by different point separations

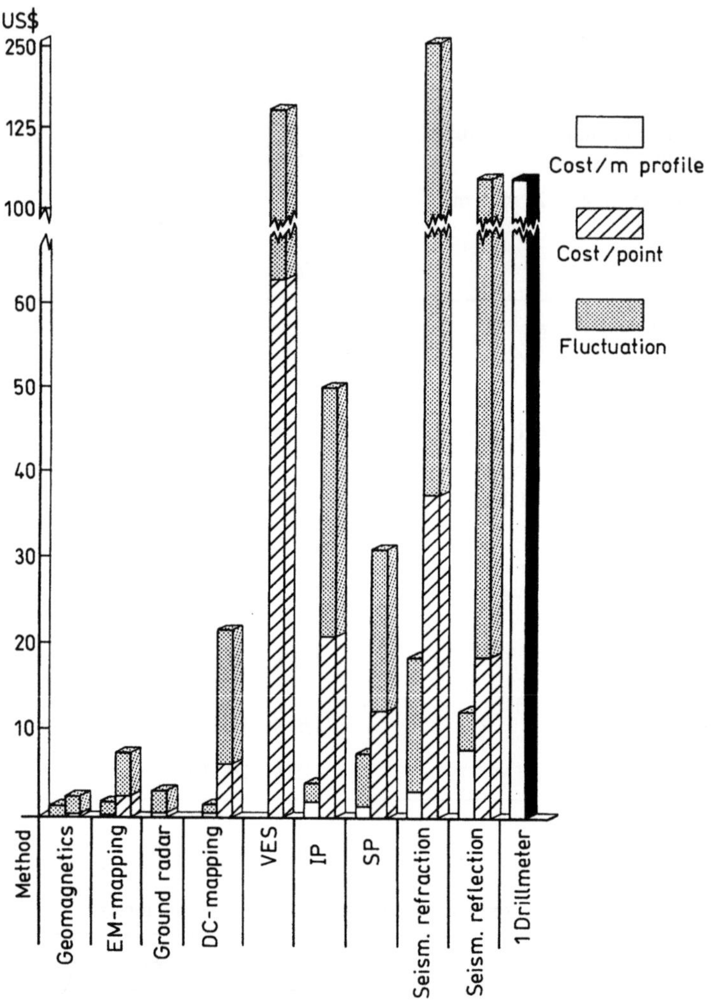

Fig. 4.1. Comparison of costs per borehole meter to expenditures for the application of geophysical methods

The expenditures of Table 4.3 were calculated for a waste site containing
domestic and industrial refuse. It allowed the complete exploration of the contaminated area of 30 000 m^2 down to a depth of 20 m. Furthermore, a distinction
between industrial and domestic waste could be made and isolated concentrations
of scrap metal, building material and industrial sludges were located.

The figures in Tables 4.1 to 4.5 concern routine arrays and spreads. For special
problems, quite different arrangements can be made. These tables must be regarded only as a rough guide to estimate the costs of an environmental investigation
before consulting a geophysical expert.

Table 4.5. Suitable point separations

Method	Point separation	Remarks
Magnetic	1–5 m	rectangular grid
Geoelectric mapping	2–15 m	rectangular grid
Geoelectric sounding	10–100 m	along paths and roads
Induced polarization	10–50 m	–
Self-potential	5–20 m	rectangular grid
EM mapping + VLF	2–25 m	rectangular grid
Ground radar	continuous	rectangular grid
Seismic refraction	1–30 m	geophone separation
Seismic reflection	5–40 m	geophone separation

5 Briefing for Geophysical Surveys

5.1 Areas of Application

Geophysical measurements or surveys can help to investigate the following:

1. Geological and hydraulically active structures of the ground under and around existing landfills or hazardous waste sites. Geological barriers for the selection of new sites.
2. Extension and contents of waste dumps. Single bodies.
3. Flow of contaminated plumes, seepages and leachates

Geophysical surveys can, in many cases, identify the stratification and the tectonic and other hydraulically active structures and textures. They are able to monitor the underground to act as a geological barrier protecting rock and ground water from putrification and contamination.

To know the exact extension of hazardous landfills and waste sites is essential for any risk assessment. In most cases, geophysics can find the borders of contaminated areas, even under thick coverage. The unexpected physical uniformity of domestic and some industrial waste furthermore allows the exploration of the underground of landfills in detail.

Geophysics can further elaborate the physical properties of buried waste, allowing the non-invasive identification of contaminated materials from above. The pursuit of the underground paths of contaminated ground water is another valuable field of geophysical exploration. This depends, however, on significant alterations of the physical properties of the ground.

Since most leachates and seepages from landfills and dumps are salty, many plumes carrying freights of soluble contaminants can be detected by geophysics. But identification of concentrations of organic fluids, e.g. fluorhydrocarbons, may fail.

Another positive effect of geophysical measurements on hazardous sites is the industrial safety of working personnel. Since geophysics is non-invasive and non-destructive, natural and artificial sealings remain intact. Toxic gases or liquids cannot escape and the health hazard is low.

Table 5.1 shows a brief overview of the most important areas of application for approved geophysical methods. Methods not mentioned, like gravity or geo-thermy, are rarely employed. The well-logging methods described intensively in this book should always be applied to any sunken borehole.

Table 5.1. Applications of Environmental Geophysics

Methods	Area of application			
	Geology	Landfill	Plumes	Remarks
Magnetic	(+)	+	–	ltd. depth penetration
Geoelectric mapping	(+)	+	(+)	disturbed by pipes etc.
Geoelectric sounding	+	+	+	disturbed by pipes etc.
Induced polarization	(+)	(+)	(+)	research necessary
Self-potential	(+)	(+)	(+)	research necessary
Electromagnetic methods	+	+	+	EM, VLF, TDEM
Ground radar	(+)	+	(+)	dry ground necessary
Seismic refraction	+	(+)	–	–
Seismic reflection	+	–	–	expensive

+ applicable, (+) limited applicable, – not applicable

The successful and cost-saving application of geophysics depends very much on the physical properties of the site and on the skill of the geophysicist in charge to combine the right methods to extract an optimum of information. It is futile to apply geophysics if the differences among physical properties are too small. Good examples are the specific electric resistivity and the seismic velocity. The resistivities of two materials to be distinguished should differ at least by 30 Ωm; seismic velocities should vary by 200 m/s.

The disturbing influence of human installations like metal pipes or cables, roads, buildings, etc. must be considered before any measuring begins. Their presence may limit any geophysical activities or demand the application of methods which remain undisturbed by installations. It is therefore recommended to first go over the area with a cable- or pipe-detector. Sometimes man-made effects are invisible: the induction of high-voltage power lines will effect electromagnetic measurements up to several hundred meters away.

5.2 Objectives and Limitations

Table 5.2 lists targets of environmental geophysics and the proper methods to be used. Only surface measurements are included.

Below, the suitability of geophysical methods to solve environmental problems is discussed in detail:

Magnetics

Magnetic methods are best adapted to the boundary mapping of domestic waste sites by the general content of iron objects. On landfills without magnetic material, magnetics will naturally fail. Single iron objects like metal barrels or car scrap can be located only near the surface. If buried deeper into the waste, they cannot be found.

Table 5.2. Objectives of Environmental Geophysics

Methods	Objectives					
	Domestic waste	Industrial waste	Abandoned site	Plumes in pore aquif.	Plumes in joint aquif.	Geological barrier
Magnetic	+	+	+	−	−	−
Geoelectric mapping	+	+	+	+	+	+
Geoelectric sounding	+	+	−	+	+	+
Induced polarization	+	+	*	+	*	+
Self-potential	*	+	−	*	−	−
Electromagnetic: EM, VLF, TDEM	+	+	+	+	+	*
Ground radar	*	*	+	−	−	*
Seismic refraction	*	*	−	+	−	+
Seismic reflection	−	−	−	+	−	*

Suitability: + = good, * = limited, − = not possible

Even in landfills of soil and excavated earth, magnetics may succeed if the magnetic properties of the deposited material differ sufficiently from the country rock.

Magnetic measurements are prone to the magnetic influence of iron installations like cast pipes, fences or casings of boreholes, and most of all, of iron buttons or buckles of the operator; care has to be taken to avoid them.

In most cases, only statements about the location and the depth of objects can be made. Identification of type, like drum, fridge or car part, is not possible.

Geoelectrics

Geoelectric measurements are restricted if buildings, metallic pipes and cables are present in the ground, or if railway lines or guard rails cross the area of investigation.

The grounding of electrodes becomes difficult on tar and concrete road surfaces. Plastic liners may be perforated.

Good conducting beds like clay and marl decrease the depth penetration.

DC measurements are time consuming, since current and potential electrodes have to be grounded.

The precondition of +/− horizontal stratification must be fulfilled for geoelectric soundings.

Geoelectric soundings may omit thin layers. Several equivalent solutions as to the depth or the thickness of beds are possible.

Such limitations are also valid for the *Induced Polarization*. In addition, electrical noise near industrial areas will disturb and distort the measurements.

The results of *self-potential* surveys are difficult to interprete since redox and flow potentials overlap and the spatial position of the plus and minus poles, or the exact location of the SP sources, are rarely determinable.

Electromagnetic measurements are disturbed by all metallic installations, and especially by powerlines running near the survey area. The depth penetration of VLF observations is limited mostly to < 20 m. VLF profiles should run perpendicular to a straight line between the object of investigation and the very distant transmitting station.

Ground radar penetrates from decimeters to a few meters into the ground. The obtained data change rapidly after rain or other alterations of soil moisture. Dense vegetation or rough ground prevents radar work. The evaluation has to compile tremendous amounts of digital data. Radar data must be cautiously interpreted. This is not possible without practical experience. If this is lacking, overinterpretations may occur.

Seismics

Seismic surveys are the most voluminous and expensive geophysical methods. They are confined to horizontal stratification. Steep structures can only be indirectly inferred from interruptions of horizontal bedding. Areas of high ground disturbance, caused by railways and road traffic, may be unsuitable for seismic surveys.

Seismic refraction is well adapted to map the surface of a consolidated bed, even below landfills or hazardous waste sites to trace the flow of leachates or contaminated ground water. The number of layers to be investigated is limited. A precondition is the increase in seismic velocity with growing depth.

The determination of the borders of waste dumps depends on the differences in the seismic velocities between refuse and country rock. If the rock is unconsolidated, the velocities may be equal and the borders not discernible.

Seismic reflection has only recently been developed to reveal structures above the depth of 50 m. Most units work only at greater depths. It can precisely disclose many horizontal boundaries, but very thin layers may be overlooked. Before seismic reflection is employed, its high costs must be considered.

Gravity

The gravity anomalies of landfills are generally very small. They can fall well below the level of necessary corrections. It is nevertheless a good method to find cavities or abandoned adits. Any environmental application demands very narrow grids, which are expensive to survey. The costs of gravity evaluations may be higher than those of fieldwork. Very small anomalies may lead to false interpretations.

Geothermy

Geothermal measurements depend on strong heat production rates inside a landfill and dense heat flows on the surface. Observations should be made only in short probing holes between 2 and 5 o'clock a. m. and in dry weather. They might be influenced by alterations of the local microclimate.

5.3 Planning and Execution of Investigations

5.3.1 Choice of Methods

Table 5.3 lists the most common problems and the routine geophysical methods of solving them. The suitability is graded from good to limited to impossible. Table 5.3 is strongly simplified for the use of geophysical laymen.

Table 5.3 should always be used in connection with Tables 5.1 and 5.2 to select the best methods or combination of methods to solve environmental problems. It must be emphasized that the use of these tables can certainly not replace consultation with a geophysicist; they are merely to be used as general briefings.

The following examples are to aid in the choice of methods: Simple problems, like finding the location and extension of a landfill or waste dump, should be tackled by the cheap and fast methods of magnetics and geoelectrics. Magnetic measurements can trace the extension of most buried domestic waste bodies; in earth dumps, magnetics can locate single iron objects.

Table 5.3. Choice of Methods for Hazardous Waste Sites

Methods	Objects to be checked							
	Locating	Extension	Top sealing	Bottom sealing	Thickness of deposit	Single objects	Seepage paths	Saline plumes
Magnetic	+	+	−	−	−	+	−	−
Geoelectric mapping	+	+	+	−	−	+	+	+
Geoelectric sounding	*	*	+	+	+	−	*	+
Induced polarization	*	*	−	−	*	*	*	*
Self-potential	*	*	−	−	−	*	*	*
Electromagnetic: EM, VLF, TDEM	+	+	*	*	+	+	*	+
Ground radar	*	+	+	−	−	+	−	−
Seismic refraction	*	*	−	+	+	−	+	+
Seismic reflection	−	−	−	*	*	−	*	+

Suitability: + = good, * = limited, − = not possible

Electromagnetic or geoelectric (DC) mapping can complement the magnetic data where no iron is present or where the depth of deposited material has to be determined. Geoelectric sounding can help to find the thickness of a dump and the channels that guide the flow of leachates.

Furthermore, geoelectric sounding and/or seismic refraction are capable of answering geohydrological questions about the thickness and extensions of aquifers, aquicludes and their overburden. In hard rock areas, fissured seepage paths and permeable faults may be traced by electromagnetic methods. The thickness of unconsolidated sediments on top of hard rock or the morphology of the interlayer surface can be found by seismic refraction.

It has to be considered whether combinations of methods should be applied simultaneously or in succession. It is advantageous to divide the survey into two phases:

Phase 1: Presurvey and overview. Only the cheap and fast magnetic, geoelectric and electromagnetic methods should be used.

Phase 2: Main survey. Areas selected by results of the presurvey are to be investigated in detail. This comprises the tightening and extension of the presurvey grid and the employment of methods like seismic refraction, seismic reflection or induced polarization.

Interdisciplinary cooperation with geohydrologists, environmental experts or the owners of waste sites is often successful. In any case, a cost-benefit analysis, based on Tables 4.1 to 4.2, is necessary to establish the best combination of geophysics, drilling and probing.

5.3.2 Necessary Experience

The performance of geophysical surveys cannot follow direct schemes and standards. The investigations have to be adapted to the special conditions of every locality. The sequence of geophysical measurements has to be arranged according to the specific results of each method, the geological background, presurvey analyses and drilling data. Therefore, the interpretation of geophysical field observations has to be based on long-term experience and should be counterchecked with environmental engineers, geologists or geohydrologists.

The choice of arrays to be combined also requires considerable expertise and knowledge. Certain aspects of environmental hazards must be followed up by different arrays. If, for example, the precise location of the margin of a hazardous waste deposit has to be known, geoelectric and/or magnetic measurements have to be made in a narrow rectangular grid of 1×1 m^2. However, the general position of buried waste deposits can be explored by wider grids of 10×10 m^2.

One should keep in mind that little and imprecise preinformation raises the cost of geophysical investigations considerably and hampers the interpretation.

The geophysical monosurvey or the application of only one method should be the exception. Combinations are obligatory to arrive at a correct solution.

Equivalent results can thus be sorted out. Naturally, not only geophysical but also geological, drilling, or technical data will assist in achieving a correct and safer result. Further information about the combination of methods is found in Sect. 5.5.

5.3.3 Preparations

The first step in planning a geophysical survey is to collect all available physical and historical data about the target. The problems and demands of the client must be defined in detail before a first blueprint can be drawn.

Knowledge of the target should include information about accessibility, distance to buildings, railway lines, roads or electrical installations, topography and, most important, the location of pipes and cables.

Most urgent is the availability of maps, sections or other graphic documents describing the status of the area. They must be scaled to enable the accurate entering of measured stations. Only precise and correct maps can serve as a topographical base for the evaluation and interpretation of geophysical data.

In some areas, digital data banks inform about agricultural use, technical installations, drilling activities, etc. They should be consulted in connection with the following Table 5.4.

5.3.4 Evaluation and Interpretation

Evaluation and interpretation have to be done with the same caution and care as the geophysical fieldwork.

Nowadays, evaluation is much easier and faster than in the past, since special geophysical software is available. Putting the computer to work saves time but does not obviate thinking. The formal and generalized computer results have to be corrected according to the specifications of the order and to the known properties of the object.

The final report has to be comprehensible to non-geophysicists, as well, but data must be so thoroughly documented that reevaluations and reinterpretations are possible. All maps, sections and other graphics should be presented in the same scale to enable direct comparison (see Sect. 1.4).

Geophysical statements as to the extension of contaminations, the hydraulic paths of seepages and the geology of an area should be discussed in detail in a final report. A mere formal presentation of geophysical data without discussing the special relevance of results is not sufficient!

5.3.5 Follow-up Activities

A scheduled geophysical survey may have to be amended during and after fieldwork. If anomalies are recognized at the ends of lines, they should immediately be prolonged. Such extensions should not be postponed until after the final evaluation, to avoid the necessary return of the field crew to the area of investigation.

Table 5.4. Checklist Preparation of Geophysical Surveys

1. Definition of the object of investigation and the problems to be solved.

2. Assessment of the topography. Possibility to drive or to walk into the area, installations such as power lines, buried metallic pipes or cables, distances to buildings, roads and railway lines.

3. Collection of information from government offices such as topographical or geological surveys, mining or environmental bodies. Records and reports about time and nature of dumped material and fitting of sealings have to be studied.

4. Assessment of geophysical methods to answer the special questions of the client, considering the information under 1. to 3.

5. Setting up a survey programme, including the positioning of the survey lines and points in a correct map. The distances between points and lines have to be chosen according to the demands for vertical and lateral penetration.

Table 5.5. Checklist for Evaluation and Interpretation

1. Application of approved software.

2. Entering the results of different methods into maps, sections and graphics on equal scale.

3. Clear presentation of results in words and graphics which is understandable also for geophysical laymen. Use of colour to enhance this.

4. Critical revision of geophysical results by comparison of the data of all geophysical methods used.

5. Comparison of geophysical data to geology, geohydrology, tectonical structures and properties of the waste e.g. to chemical analysis.

Table 5.6. Checklist Follow-up Work

1. Execution of complementary field measurements to consolidate the results.

2. Extension of the survey into adjacent prospective areas, e.g. to trace contaminated plumes.

3. Verification of geophysical results by drilling, probing or trenching programmes.

Modern digital field instruments can store the measured data of one day or more. Transferred to laptops directly in the field, evaluation software can be applied to check the daily output for faulty measurements. Questionable data must be repeated or counterchecked in the field the following day.

In many cases, a wide-meshed grid is laid first to recognize areas with characteristic indications. If, however, such anomalies are found by a preliminary evaluation, those areas have to be surveyed in a second field campaign within narrow-meshed grids (Table 5.6). This confinement of detailed surveying to prospective areas will always result in considerable cost reduction.

Follow-up work should lead ultimately to the positioning and realization of drilling, probing and trenching programs. Since these are very expensive, great care has to be taken by the geophysicist in charge to recommend such developments only after a careful scrutiny of the geophysical results.

5.4 Combination of Geophysical Methods

Every geophysical method is directed at certain physical properties of material or rock. It is advisable not to investigate single properties by geophysical methods, but as many properties as possible by various methods, to countercheck and consolidate the results.

Methods to be combined are described in Table 5.7, which may be considered an extension of Table 5.2. Nevertheless, these tables contain only general advice. The combination of geophysical methods for individual objects can be quite different, depending on the aim of the survey and on special local conditions.

Experience has shown that at most waste sites or landfills, the cheap and fast methods of magnetics and geoelectric or electromagnetic mapping suffice to determine the spatial extension. If, however, single waste components have to be pinpointed, more efforts and more methods are required. Ground radar, induced polarization, self-potential or seismic refraction may have to be used.

The costly seismic reflection is to be recommended when information from a greater depth is desired. This might be necessary if contaminated plumes plunge into deep strata.

Ground radar is the only method that is influenced by the dielectric constant. Since this is very different for water and organic fluids, it is a good method to trace strong organic contaminations of ground water. Furthermore, it is capable of detecting non-metallic pipes, which cannot be found by other methods. Because of its very limited depth penetration, it must often be supplemented by deeper-reaching methods, such as electromagnetic mapping.

It is very important to run geophysical logs in every borehole. The logs record very precisely the alterations of physical properties in vertical extension. Good geophysical borehole data will upgrade the information from surface geophysics many times!

5.5 Research and Development

Environmental geophysics is a very young applied science, and employs mostly the procedures of mining geophysics adapted to shallow targets by smaller arrays and spreads. This was adequate at the beginning, but to develop this new branch of geosciences, further research work is necessary.

One drawback to environmental geophysics is the lack of knowledge of the physical properties of waste. Sometimes they are quite different from what might

Table 5.7. Combinations of Geophysical Methods

Target	Methods						
Landfills, hazardous waste dumps	magnetics	geoelectrical mapping/ sounding	EM mapping TDEM, VLF	ground radar	induced polarization	self-potential	seismic refraction
Abandoned industrial sites or areas	magnetics	geoelectrical mapping/ sounding	EM mapping	ground radar		self-potential	
Contamined plumes, seepage paths		geoelectrical mapping/ sounding	EM mapping TDEM		induced polarization		seismic refraction
Geological barrier		geoelectrical mapping/ sounding	EM mapping VLF		induced polarization		seismic refraction seismic reflection

be expected. Who would have thought that the very inhomogeneous domestic waste behaves geoelectrically like a homogeneous body of very low resistivity?

It is necessary that the tables in this book, which list some physical properties known from individual determinations, shall be amended by laboratory and field experiments. Even this enhanced knowledge of physical properties is not enough; the specific responses of waste materials and fluids to geophysics have still to be studied.

One of many research targets is the differentiation among substances by complex resistivity or induced polarization measurements, using the mathematical tool of the fractal dimension. This new geophysical approach should be used to distinguish between clays and unconsolidated rock filled with saline ground water.

Self-potentials may originate from oxidation and reduction, or from the movement of fluids through permeable substances. Research work should be directed toward finding a means of telling the difference between the two effects.

High-frequency seismics are now able to work in subsurface areas. But interpretation is still difficult, since very few seismic velocities of human refuse are really known.

A new field of research concerns the long-term monitoring of leaks in bottom seals. If a leak occurs, it can be detected only much later, after its seepage has reached one of the monitoring wells outside the landfill. This may take weeks or months and may result in high remedial expenses.

Tests arranged by perforating the plastic liner (Fig. 5.1) have indicated spreading of the artificial plume by 3.6 m after 24 h and 45 mm of rainfall, and by 5.3 m after 96 h and 62 mm of rainfall. The water seeped away below the plastic liner and collected on top of the clay barrier. It would have taken about 40 days until

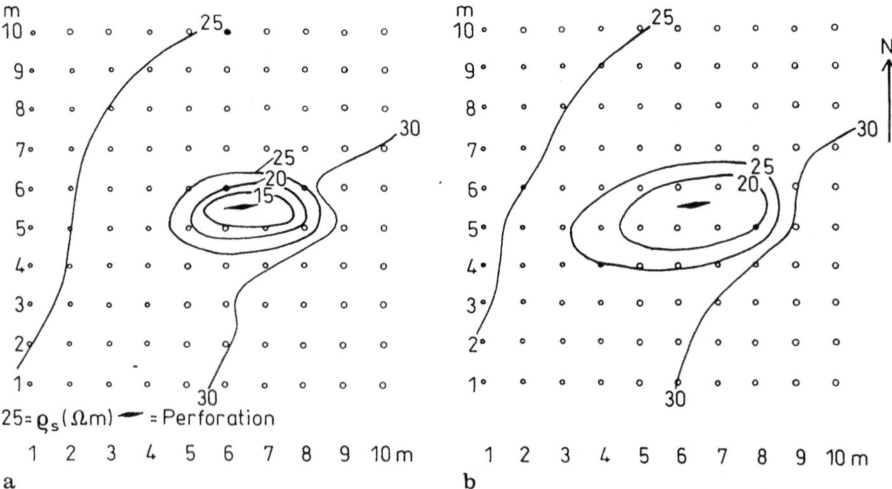

Fig. 5.1 a, b. Alteration of the electric resistivity of a bottom clay barrier after perforation of the plastic liner. **a)** after 24 hours and 45 mm rain; **b)** after 96 hours and 62 mm rain

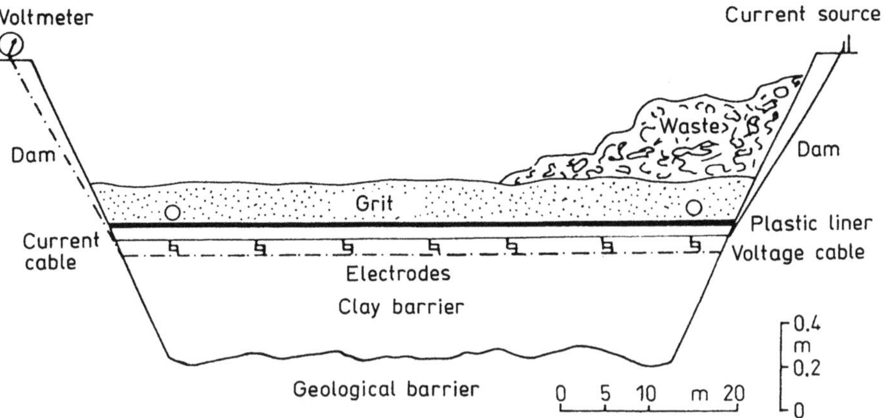

Fig. 5.2. Principle of long-term leak monitoring by permanent electrodes

this artificial plume had reached monitoring wells at the margin of this landfill of 150×350 m^2.

To avoid this crucial delay, it is proposed to implant stainless steel electrodes into the clay barrier, directly below the plastic liner. By connecting them up by rustproof cables, it is possible to monitor their electric potential in a common array in short time intervals (Fig. 5.2). If leachates, especially saline leachates, seep between two electrodes, their potential will abruptly diminish and the DC current will change. This monitoring system will report the time and the location of leaks immediately. Remedial measures could thereby commence at once.

In a pilot project, the stainless steel electrodes should be separated by 10-m intervals. They should be round with a diameter of 2.5 cm and arrayed in a rectangular grid. Their connectors have to be riveted to non-corrodable cables that lead to a constant power source and a digital voltmeter with automatic memory functions. The shunting of different arrays, preferably the Wenner array, must be possible.

Other research work should answer the question of whether long-term application of electric potentials or DC currents to aquifers might help in the precipitation of toxic ions. The experiences already accumulated in preventing pipe corrosion by weak electric potentials could be used.

These are only a few selected examples of urgent research activities in the field of environmental geophysics. With the growing application of geophysics to shallow environmental targets, new problems will have to be solved, necessitating more research and development. Geophysics should also be employed to aid and to check remedial or mitigation programs.

6 Geophysical Tenders

6.1 Procedures

After the combination of methods has been chosen, the call for tenders can be made. Table 6.1 lists important information for bidders.

Tenders should be kept flexible. Mainly, the total volume of geophysical work has to be described, e.g. the number of points or stations, size of the area to be surveyed, etc. The bidder should be able to include his own ideas concerning the array, the point and line separations.

Details must be fixed only if necessary to fulfill the purpose of survey. However, requirements about the requested precision and depth penetration are indispensable. For some methods, certain geophysical instruments have to be contracted (e.g. the proton magnetometer).

It has to be stipulated how the final data are to be presented. This might be in cross sections, profiles, or three-dimensional pictures, colored or black and white. Any reports must be accurate, understandable, and meet the current standards of science and technology. Further demands concerning the presentation should not be made in tenders, since many details of the presentation can only be decided upon after the field data have been collected.

Table 6.1. Information Recommended for Bidders

Position, size and topography of survey area	Topo maps with proposals for survey areas, arrays	Number of survey points, length of profiles, grids
Details of object and problems	Details of geology, hydrology, tectonics and dumped materials	Time table: Beginning and end of field work and evaluation

Table 6.2. Compilation of Tenders and Placement

1. Compilation of an appropriate flexible tender, with precise description of the object, work program, fieldwork, evaluation, interpretation and reporting.

2. The prices for preparation of equipment, transport, field measurements, stand-by times, digital evaluation, interpretation and presentation/reporting have to be declared separately.

3. Placing the order to the consultant who made the best bid at acceptable total costs, and with good references as to his reliability and dependability.

The evaluation and interpretation of geophysical data can be quite extensive and time consuming, since indirect conclusions must be worked out. These expenses might eventually surpass the costs of the field operations. Therefore, the field and evaluation costs in bids have to be separated.

Very important is the compatibility of the results of different geophysical methods. To achieve this, measuring points of all methods have to be entered into maps on equal scales.

It might not be wise to contract the cheapest bidder. The one with the most relevant experience (Table 6.2) and who can submit references as to his dependability and proper fulfillment of previous contracts should be chosen.

6.2 Call for Tenders

6.2.1 Preparation

Information for the bidder:

– Method or combination of methods.
– Size (arrays) of survey grid, point separation.
– Instruments to be employed.

Requested special prices:

– Topographical surveying and pegging .

6.2.2 Fieldwork

Requested Prices:

– per measured point.
– per line meter (kilometer etc.).
– for additional measures (e.g. monitoring diurnal field variations, error margins).

6.2.3 Evaluation

Requested Prices

– per corrections (topographical, diurnal).
– per application of sophisticated software (e.g seismic migration, geoelectric inversion).
– per 2D and 3D derivations.
– per digital model calculations.
– per technical or geological interpretations.
– per reporting, including three-dimensional and/or colored pictures, if necessary.

7 List and origin of Figures and Tables

7.1 List of Figures ([ ] = No. of references)

7.2 List of Tables

8 References

1. Barker, R. D., 1990: Investigation of ground water salinity by geophysical methods. – In Ward, Geotechnical and Environmental Geophysics (Publ.), ISBN 1-56080-001-1 II, 201–213; Tulsa.
2. Berktold, A.; Schleicher, F.; Strobel, P.; Mathes, P.; Durlesser, H. P., 1992: Möglichkeiten und Grenzen des VLF-R Verfahrens im Ingenieur/Umweltbereich. – Geophys. Mitt. 1, 65–86, ISSN 0931-2145; Munich.
3. Bisdorf, R. J., 1990: Geoelectrical studies of the Panoche fan area of the San Joaquin Valley, California. – USGS Circ. 1033-133-139; Denver.
4. Boom-Van Den, G. & Ort, Matthias, 1991: Anwendung eines gasgeochemischen Reconnaissance Verfahrens zur Bestimmung von Helium und Radon in der Bodenluft. – Report BGR-Archiv Nr. 108203, 1–23; Hannover.
5. Bredewout, J.W., 1990: Detection of iron objects with magnetic and EM methods. – 52 EAEG Meeting; Copenhagen.
6. Brost, E., Ellwanger, D., 1991: Ergebnisse neuerer geoelektrischer und stratigraphischer Untersuchungen im Gebiet zwischen Kaiserstuhl und Kehl. – Geol. Jb. E 48, 71–81, ISSN 0341-6410; Hannover.
7. Buchholz, R.: Unpublished reports, figures and drawings; Heiligenberg.
8. Bundesanstalt für Geowissenschaften und Rohstoffe (Hrsg.), 1987: Die Erde. – Erforschung zum Nutzen der Menschen. Hannover.
9. Bundesminister für Forschung und Technologie (Hrsg)., 1992: Verbundvorhaben „Methoden zur Erkundung und Beschreibung des Untergrundes von Deponien und Altlasten". – BGR-Archiv Nr.: 109–492; Hannover.
10. Buseli, C., Barber, C., Davies G. B. & Salama R. B.,1990: Detection of ground water contamination near waste disposal sites with transient electromagnetic and electrical methods. – In Ward, Geotechnical and Environmental Geophysics (Publ.), ISBN 1-56080-001-1 II, 27–41; Tulsa.
11. Butler, D. K. & Llopis, J. L.,1990: Assessment of anomalous seepage conditions. – In Ward, Geotechnical and Environmental Geophysics (Publ.), ISBN 1-56080-001-1 II, 153–175.
12. Daniels, J.L. & Keys, W.S., 1990: Geophysical well logging for evaluating hazardous waste sites. – In Ward, Geotechnical and Environmental Geophysics (Publ.), ISBN 1-56080-003 I, 263–287; Tulsa.
13. Darilek, G. T. & Laine, D., 1989: Understanding electrical leak loacation surveys of geomembrane liners and avoiding specification pitfalls. – Proc. 10th Nat. Conf. Superfund, 56–67; Silver Spring.
14. Duerbaum, H.-J.; Reichert, C.; Bram, K., 1990: Integrated Seismics Oberpfalz 1989. – KTB-Report, 90–96b; Hannover.
15. Duval, J.S. & Tanner, A.B., 1990: Some basic facts about radioactive radon. – USGS Circ. 1033, 163–165; Denver.
16. Environmental Protection Agency: Geophysical techniques for sensing buried wastes and waste migration. – EPA/EMSL, EPAX 8706-0050; Las Vegas.
17. Foerstner, U., 1990: Umweltschutztechnik. – 1–462, Springer, ISBN 3-540-52154-2; Berlin.

18. Fraser, D., 1978: Resistivity mapping with an airborne multi-coil electromagnetic system. – Geophysics 43 144–172.
19. Frischknecht, F.C., 1990: Application of geophysical methods to the study of pollution associated with abandones and injection wells. – USGS Proc. Circ. 1033 73–79; Denver.
20. Geyh, M. A., 1988: Methoden der Umweltisotope – In Schneider, H. (Hrsg.) Die Wassererschließung, 339–354; Essen.
21. Geophysik Consulting GmbH: – Unpublished reports, figures and drawings; Kiel.
22. Goldsten, N. E. & Benson, S. M., 1990: Saline ground water plume mapping with electromagnetics. – In Ward, Geotechnical and Environmental Geophysics (Publ.), ISBN 1-56080-001-1 II, 17–27; Tulsa.
23. Grüneberg, S., 1991: Unpublished drawings; Hannover.
24. Hagemeyer, R. T. & Steward, M., 1990: Resistivity investigations of salt-water intrusions near a major sealevel canal. – In Ward, Geotechnical and Environmental Geophysics (Publ.), ISBN 1-56080-001-1 II, 67–77; Tulsa.
25. Hasbrouck, W.P., 1990: Initial shallow seismic tests in the Panoche fan area, Fresno County, California. – USGS Circ. 1033 139-143; Denver.
26. Hering, E., Schulz, W., 1987: Kernkraftwerke, Radioaktivität und Strahlenwirkung. – VDI-Verlag 1987, ISBN 3-18-400793-6; Düsseldorf.
27. Hoekstra, P. & Blohm, M. W., 1990: Case histories of timedomain electromagnetic soundings in environmental geophysics. – In Ward, Geotechnical and Environmental Geophysics (Publ.), ISBN 1-56080-001-1 II, 1–17; Tulsa.
28. Homilius, J.; Flathe, H., 1988: Geoelektrik in der Wassererschließung. In: SCHNEIDER, H. (Hrsg.): Die Wassererschließung, 3. Aufl. Vulkan; Essen.
29. Howard, K. W. F.,1990: The role of well logging in contaminated transport studies. – In Ward, Geotechnical and Environmental Geophysics (Publ.), ISBN 1-56080-001-1 II, 289–301; Tulsa.
30. Lankston, R. W., 1990: High-resolution refraction seismic data acquisition and interpretation. – In Ward, Geotechnical and Environmental Geophysics (Publ.), ISBN 1-56080-003 I, 45–75; Tulsa.
31. McNeill, J. D., 1990: Use of electromagentic methods for ground water studies. – In Ward, Geotechnical and Environmental Geophysics (Publ.), ISBN 1-56080-003 I, 191–219; Tulsa.
32. Militzer, H.; Weber, F. (Hrsg.), 1985: Angewandte Geophysik. – Springer; Wien, New York.
33. Mundry, E. & Homilius, J., 1979: Three-layer model curves for geoelectrical resistivity measurements – Schlumberger array. – Schweizerbart; Stuttgart.
34. Nielson, D. L., Linpei, C. & Ward, S. H., 1990: Gamma-ray spectrometry and radon emanometry in environmental geophysics – In Ward, Geotechnical and Environmental Geophysics (Publ.), ISBN 1-56080-003 I, 219–250; Tulsa.
35. Olhoeft, G. R., 1986: Direct detection of hydrocarbon and organic chemicals with ground-penetrating radar and complex resistivity. – Proc. NWWA/API Conf. & Natl. Water Well Ass. 284–305; Houston.
36. Repsold, H., 1989: Well Logging in Ground water Development. – Int. Ass. Hydrogeol. 9, 1–136: ISSN 0936-3912; Hannover.
37. Richardson, W. K., Kirkpatrick, G.L. & Cline, S. P., 1989: Integration of borehole geophysics and aquifer testing to define a fractured bedrock hydrogeological system. – Proc. 10th Nat. Conf. Superfund, 277–282; Silver Spring.
38. Roberts, R. L., Hinze, W. J. & Lep, D. I., 1990 (1): Application of the gravity method to the investigation of a landfill in glaciated midcontinent USA. – In Ward, Geotechnical and Environmental Geophysics (Publ.), ISBN 1-56080-001-1 II, 253–261; Tulsa.
39. Roberts, R. L., Hinze, W. J. & Lep, D. I., 1990 (2): Data enhancement procedures on magnetic data from landfill investigations. – In Ward, Geotechnical and Environmental Geophysics (Publ.), ISBN 1-56080-001-1 II, 261–267; Tulsa.

40. Ross, H. P., Mackelprang, C. E. & Wright T, P. M., 1990: Dipole-dipole electrical resistivity surveys at waste disposal study sites. – In Ward, Geotechnical and Environmental Geophysics (Publ.), ISBN 1-56080-001-1 II, 145–153; Tulsa.

41. Rueter, H.; Elsen, R., 1989: Geophysikalische Methoden bei der Altlastenerkundung.- Thome'-Kozmiensky, K. J. (Hrsg.) 3, 89–115; Berlin.

42. Sengpiel, K. P., 1983: Resistivity/depth mapping with airborne electromagnetic survey data. Geophysics 48 181–196.

43. Slaine, D. D., Pehme, P. E., Hunter, J. A., Pullan, S. E. & Greenhouse, J. P., 1990: Mapping overburden stratigraphy at a proposed hazardous waste facility using shallow seismic reflection methods. – In Ward, Geotechnical and Environmental Geophysics (Publ.), ISBN 1-56080-001-1 II, 273–280; Tulsa.

44. Steeples, D. W. & Miller, R. D., 1990: Seismic reflection methods applied to engineering, environmental and ground water problems.- In Ward, Geotechnical and Environmental Geophysics (Publ.), ISBN 1-56080-003 I, 1–31; Tulsa.

45. Struttmann, T. P. E. & Anderson, T., 1989: A comparison of shallow electromagnetic (EM31) and proton magnetometer surface geophysical techniques to effectively delineate pre-RCRA buried wastes.- Proc. 10th Nat. Conf. Superfund, 27–35; Silver Spring.

46. Thierbach, R. & Mayerhofer, H., 1978: Elektromagnetische Reflexionsmessungen in Salzlagerstätten.- Fifth international symposium on salt, 393–403; Hamburg.

47. Thor Geophysikalische Prospektion GmbH: Unpublished reports, drawings and figures; Kiel.

48. Van Zijl, J. S. V.; Köstlin, E. O., 1985: The electromagnetic method.- Field man. f. technicians. S. A Geophys. Ass; Joahnnisburg.

49. Vogelsang, D., 1988: Bohrlochmessungen in Bohrungen zur Vorerkundung der KTB-Lokationen Oberpfalz und Schwarzwald,Geol. Jb. E 41 : 5–19; Hannover.

50. Vogelsang, D. & Stettner, G., 1988: Bohrungen zur Vorerkundung der KTB-Tiefbohrlokation Oberpfalz - Petrographie und Bohrlochgeophysik.- Geol. Jb. E 41 : 27–42; Hannover.

51. Vogelsang, D.; Schimer, W.; Strassburger, A., 1990: Materialien zur Altlastenbearbeitung.- Leitl.Geophysik an Altlasten LFU, 2, 1–137; Karlsruhe.

52. Vogelsang, D., 1991: Geophysik an Altlasten, Leitfaden für Ingenieure, Naturwissenschaftler und Juristen. – Springer 1–133; ISBN 3-540-53948-4; Berlin, Heidelberg.

53. Vogelsang, D., 1993: Geophysik an Altlasten, Leitfaden für Ingenieure, Naturwissenschaftler und Juristen. – Springer, 2nd enlarged Edition, 1–179, ISBN 3-540-56659-7; Berlin, Heidelberg.

54. Wallach, G., 1991: Erkundungsmethodische Grundlagen der Risikobewertung von Altlastverdachtsflächen. – Radex-Rdsch 3/4, 583–597; Leoben.

55. Ward S. H., 1990: Resistivity and induced polarization methods.- In Ward, Geotechnical and Environmental Geophysics (Publ.), ISBN 1-56080-003 I, 147–191; Tulsa.

9 Index

Legend:
120 = page number
f = >1 word per page
( ) = adjective or synonym
→ = found under

Springer-Verlag
and the Environment

We at Springer-Verlag firmly believe that an international science publisher has a special obligation to the environment, and our corporate policies consistently reflect this conviction.

We also expect our business partners – paper mills, printers, packaging manufacturers, etc. – to commit themselves to using environmentally friendly materials and production processes.

The paper in this book is made from low- or no-chlorine pulp and is acid free, in conformance with international standards for paper permanency.

If you have any concerns about our products,
you can contact us on
ProductSafety@springernature.com

In case Publisher is established outside the EU,
the EU authorized representative is:
Springer Nature Customer Service Center GmbH
Europaplatz 3, 69115 Heidelberg, Germany

Printed by Libri Plureos GmbH
in Hamburg, Germany